ON THE DISTRIBUTION OF THE MOON'S HEAT.

PRIZE ESSAY

ON

THE DISTRIBUTION

OF THE

MOON'S HEAT

AND

ITS VARIATION WITH THE PHASE

BY

FRANK W. VERY
of the Allegheny Observatory, U. S.

SPRINGER-SCIENCE+BUSINESS MEDIA, B.V. 1891

ISBN 978-94-017-5706-5   ISBN 978-94-017-6042-3 (eBook)
DOI 10.1007/978-94-017-6042-3

This Essay was presented to the UTRECHT SOCIETY OF ARTS AND SCIENCES, in response to the Proposition:

„On demande de déterminer la chaleur donnée par la Lune dans des phases diverses."

It obtained the Prize in the General Assembly of the Society, held at Utrecht on the 2$^\text{d}$ July 1890.

CONTENTS.

On the Distribution of the

MOON'S HEAT

and its Variation with the Phase.

By FRANK. W. VERY.

In the following research the problem of the variation of the moon's radiant heat with the phase is attacked by the aid of an extremely sensitive apparatus, and by a method, which is at any rate novel, and which, it is hoped, has yielded results of precision as well as of interest.

The instrument which has been relied on, for measuring the very small rise of temperature produced by the lunar rays, has been the bolometer, used in conjunction with a very sensitive galvanometer; and the plan has been to form an image of the moon, of about 3 centimeters diameter, by means of a concave silvered glass reflector of 30.3 centimeters diameter and 313.7 centimeters focal length, and to measure, not the heat from the whole of this image, but only that in a limited part of it from $\frac{1}{25}$ to $\frac{1}{30}$ of the area of the apparent disc. The observation being repeated at different points on the moon's disc, and at different phases, gives material for a series of maps showing the distribution of heat in this image, and by summation the total heat, at each of several epochs from the first to the last quarter of the moon.

INSTRUMENTS.

The instruments used in this private research are those belonging to the observatory at Allegheny, Pennsylvania, (latitude 40° 27′ 41″.6 north, longitude 5h. 20m. 02s.93 west of Greenwich) where it has been made, and the apparatus, which has been placed at the writer's private use for this purpose, will be found so fully described in the official publications of the observatory that it is only necessary here to briefly refer to it [1]).

The bolometer used in this research exposes a sensitive surface of about 19 sq. mm. It is covered by a diaphragm of white card pierced by a central circular aperture, 0.54 cm. in diameter, which permits a more accurate setting, especially in the condition of tangency to the moon's limb. The part of the image of the moon selected for measurement is brought to coincide with this aperture by the adjusting cords of the siderostat, all the coarser details of the surrounding lunar surface being readily seen on the white card. The adjustment having been made, the clock-work of the siderostat is trusted to keep the image fixed within small limits which will be described further on.

The siderostat-mirror, (also of glass silvered on the front surface) is large enough to give a beam which more than covers the concave mirror at all times, and the second mirror being placed centrally in the path of the horizontal reflected beam, with its optical axis slightly inclined to the direction of the former, gives an image on the white card bright enough, as has been said, to be readily seen in much of its detail, and hot enough to give, with the extremely sensitive heat measuring apparatus, a deflection of nearly 100 mm. divisions on the galvanometer scale from the small part near the center of the disc of the full moon.

The adjustment of the apparatus having been completed, a large screen of blackened copper, containing water at a known temperature, is introduced in the path of the light reflected from the siderostat, but at a considerable distance from the concave mirror. This screen moves between stops and can be drawn aside by the observer, while seated at the galvanometer, by pulling a cord, and restored in like manner.

MODE OF PROCEDURE.

Each observation consists in a reading first on the screen, then on the moon, and lastly on the screen again, the middle reading being compared with the mean of the first and last to eliminate any slight continuous drift of the galvanometer-needle. In addition to the measurements of the heat from the moon relatively to that from the screen, each evening's work includes one or more comparisons of the radiations from the screen and the sky, so that all

---

1) For the use of the excellent silvered glass concave mirror, I am indebted to the maker, Mr. J. A. Brashear, of Allegheny.

of the results can finally be stated with reference to the sky-radiation.

It would in some respects be preferable to let an assistant throw the image of the moon on and off the bolometer by moving the siderostat-mirror, adopting suitable precautions to prevent any influence on the instrument arising from the radiations of the heat of the body; but as this would have required the services of a trained observer, it had to be dispensed with. It is believed, however, that the method pursued is very nearly as accurate, and in any case there should be some measurements of the radiation from a body of known temperature, since the effective temperature of the sky can only be inferred in this way.

The positions on the lunar disc selected for measurement are sufficiently indicated to enable any one acquainted with the geography of the moon to recover them. Where an angle is given, it records the distance of the center of the bolometer-aperture in degrees of lunar latitude or longitude from the mean lunar equator and central meridian, but to avoid confusion the terms east and west are always used in their terrestrial signification, west being synonymous with „ preceding " and east with „ following " in the astronomical sence.

By a setting, described as „ tangent to limb ", is meant a position of the bolometer with the edge of its circular aperture distant about ½ minute of arc (or from 10° to 15° of lunar longitude or latitude) from the edge of the disc, and not therefore in geometrical tangency to it. This allowance is necessary on account of the small fluctuations in the position of the image reflected from the siderostat-mirror, which may equal, but seldom exceed this amount during the time consumed by a single measure. The same remark applies to the condition of tangency to the terminator, with the proviso that the uncertainty of position will not in general exceed 2° of lunar longitude at the center of the disc, or about one-tenth of the bolometric aperture. It should be noted also that a variation of 10° or 15° of lunar longitude or latitude in the position of the edge of the bolometer aperture at the moon's limb represents a change of only 2° or 3° in the mean place of the center of this aperture, its entire width including here about 50° of an arc of a great circle of the moon.

The libration of the moon is on the average a quantity of the same order as the uncertainty in the mean position of the bolometer, but the points chosen for measurement have been described and determined by reference to the markings on the moon's surface, and may be completely identified by reference to the map of the moon, except in the case of settings tangent to the limb, where the distances given in degrees from the great circles passing through the center of the disc, must be corrected by the amount of the libration to transform them into true lunar longitudes or latitudes.

The bolometer and galvanometer employed are capable of giving repeated measures on a source of unchanging radiation with an error less than one per cent. of the quantity to be measured.

A series of ten observations of the radiation from a boiling Leslie's cube, with the instruments used in this research, gave a mean deflection of 342.4 divisions $\pm$ 0.6 divisions, where the probable error amounts to less than 0.2 of one per cent.

The constant of a delicate galvanometer suffers change in the course of time, but these changes have been here eliminated by a method of standardizing.

The bolometer was exposed at least once on each night of observation to the radiation from a hot blackened screen of known temperature, and the galvanometer deflection, by comparison with that from a corresponding cold screen, gave the instrumental constant for that night. This galvanometer deflection per degree centigrade of the excess of temperature in the standard radiating body, when reduced to a battery current of 1/30 ampere (battery galvanometer $=$ 100 divisions), represents in its variation from night to night the change of instrumental sensitiveness, which may be due to changes in the bolometer, or in either the astaticism or the magnetic intensity of the galvanometer needles, or to other causes which it is neither necessary to specify, nor to estimate separately, since their combined effect is alone required. All of these variations have been eliminated by the mode of reduction. A standard „deflection per degree centigrade" has been adopted as the normal condition of sensitiveness, and all observations are reduced to this standard of instrumental sensitiveness, by the application of an appropriate multiplying factor, which is simply the ratio of the adopted standard deflection per degree centigrade, divided by the observed deflection per degree centigrade reduced to standard battery current.

The bolometer, while not primarily designed for absolute measurement, is nevertheless capable of having its indications transformed into the units more generally employed. In the present research, for example, one millimeter division of the galvanometer scale corresponded approximately to a radiation of about 0.00 0004 small calories received per minute on the face of the bolometer of which only about one sixth or 0.00 00007 calories was retained by the bolometer strips.

Since the amount of the moon's radiation varies with the atmospheric absorption, and this with the moon's altitude, a still further reduction to zenith is needed to make the measures, even of a single evening, comparable with each other.

In order to reduce each heat-measurement to that which might be expected with the moon in the zenith, an average apparent transmission of 0.8 for one atmosphere of 760 mm. pressure is adopted, and a multiplying factor is found by interpolation from the following table:

True air-mass Atmospheres	Transmission	Zenith Factor
1.0	0.800	1.000
1.1	0.782	1.022
1.2	0.765	1.046
1.3	0.748	1.069
1.4	0.732	1.093
1.5	0.716	1.118
1.6	0.700	1.143
1.7	0.684	1.169
1.8	0.669	1.195
1.9	0.655	1.223
2.0	0.640	1.250

GEOMETRICAL REPRESENTATION OF DISTRIBUTION OF HEAT IN LUNAR IMAGE.

We may conveniently represent the results of the observations and at the same time show the real distribution of heat at the lunar surface, indicating the fact of its increment with the altitude at which the sun is shining above that surface, and also the part which reaches the earth at any phase, by a geometrical figure, thus:

Let p = projection of the lunar diameter which is parallel to the axis of the earth's orbit, the plane of projection being perpendicular to this axis. (See Fig. 1). The circle tbs is the projection of the lunar sphere on this plane.

ap = direction of rays received from the sun.

Let θ = longitude of the limb (mp) on illuminated side from the terminator (tp). It will equally well represent the angle from the limb of any meridian selected for measurement at the full.

mn represents the projection of the moon's limb at some stage intermediate between new moon and first quarter.

ts represents projection of moon's limb at full.

The meniscoid $tasb$, produced by the revolution around an axis ap of the curves (which we will here assume to be circles having c and p for their centers) may be taken as a first approximation to a figure which will show the distribution of heat on the illumined side of the moon, and the portion of the meniscoid of revolution cut off by the plane whose trace is *lp* (projection of the illuminated limb) will represent approximately the lunar heat at the stage, when *mpt* is the illuminated part seen from the earth.

spt being the initial line, and *p*, the center of co-ordinates,

calling r θ the polar co-ordinate of point *l*

and   r$'\theta$ „   „   „   „   „   *m*

lm = r——r$'$ must be proportional to measured heat for angle θ.

Taking *ab* as unity, such a position is assumed for *c* that the variation of the quantity r—r$'$ shall nearly represent the change in the heat observed at different angles from the terminator. In this way the following rough values are obtained. For the purpose of comparison, some results of the actual measures, to be described farther on, are annexed.

θ	(If cp = $\frac{1}{10}$ bp) r—r$'$ (computed)	Observed Heat Along Equator	Observed Heat Along Meridian
0°	0	0	0
30	50	63	76
60	88	89	93
90	100	100	100

If the assumptions here made were correct, and if the observed values truly represented the distribution of heat along great circles of the moon, a compound ellipsoid, with its longer axis parallel to *tps* would be needed, instead of the larger sphere *tas*, to represent the distribution of heat over the entire surface; but actually our measuring instrument is not a mere point, but takes in a considerable area of the lunar surface; hence we must consider not only the distribution of the heat, but also the presentation of the surface.

The above is a preliminary trial to see if the results of observation can be approximately represented by such intersecting spheres as are here described (choosing *c* at such a distance from *p* as to most nearly agree with facts). The comparison shows that a more complicated figure will be needed. This is not attempted here, but instead another and yet simpler mode of representation is chosen, that of an assumed „flat moon." It is of course a pure convention. The former had some approximation to the reality. This has none, but, nevertheless, since it gives the facts of observation, and is the simplest conception possible, it is adopted.

Neglecting then, for the present, the actual areas of the portions of lunar surface visible through the aperture of the bolometer, *the facts of observation may be represented by assuming the moon to be a flat disc and drawing curves whose ordinates are equal to the heat measured along a diameter of the disc.* Such curves are given for the full moon in Fig. 2, and for the first day after the first quarter in Figs. 4 and 5; while Figs. 3 and 6 show the presentation of zones 30°, 60°, and 90°, from the points of vertical illumination at these epochs.

If the moon had no absorbent atmosphere, and if its surface were of uniform temperature and emissive power, the inequalities (i. e. hills and valleys) and the rotundity would make no difference in the radiation from different parts, for the radiation emitted from the unit of surface in any direction varies inversely as the cosine of the angle of emission, and the apparent area of the unit of surface varies in the same ratio. The disc of the moon would then *radiate* equally from all parts under these circumstances. The reflection, however, would follow a different law, because on a rough sphere, the shadows of the projections would constitute a considerable and varying part of the apparent surface presented to the observer.

The disappearance of all shadows about the time of full moon is no doubt the cause of the peculiarly large maximum in the phase-curve for light, signallized by Zöllner from his own and Herschel's observations. There is another cause tending to produce the same effect in the variation of the emitted radiation. The diminution (by foreshortening) in the relative apparent areas occupied by the colder zones of the moon at the full, is obvious from a mere comparison of figures 3 and 6, and from the consideration that a belt 10° from the terminator at the quarter phases, occupies 0.22 of the illuminated half disc, while a similar zone at the full represents only 0.03 of the projection of the hemisphere; and the consequent relative preponderance of the hot regions at the latter stage, must produce somewhat the same effect in the phase-curve for the proper radiation of the moon as that noted in the light-curve, although this effect, flowing from a different cause, will not necessarily follow the same law as that governing the light of the moon.

As Zöllner's research on the reflection by the moon indicates that its surface is everywhere roughened by elevations having an average slope of about 52°, the radiations emanating from a zone at this distance from the center should have an average angle of emission of 90°, and portions of surface on either side of this zone a diminishing angle of emission; but if the principle enunciated above be accepted, it will not be necessary to consider this complicated law, except for the small part of the lunar rays derived from reflection of the sun's, and here Zöllner's table, which is quoted on page 59, supplies the need.

DETAILED ACCOUNT OF OBSERVATIONS AND RESULTS.

The following are the observations with such astronomical data as are necessary for their reduction. Only nights of reasonably clear sky have been chosen for this work, with a single exception which is included to show the action of an amount of haze which would not interfere with ordinary measures of position, but which is fatal to accurate thermal work.

The measurement of the local distribution of heat on the moon's surface, and the summation of the separately measured portions, are essential features in the method chosen for determining the variation of the moon's heat with the phase. In fact, a measurement made with an instrument of large heat-receiving surface, of which the lunar image occupies only a portion, and this a variable portion as the moon waxes and wanes, can only give an approximate result. With such an instrument, which, so far as I know, is the only one heretofore employed, a part (x) of the heat-measuring instrument receives the varying radiation (M) from the bright part of the moon which it is desired to measure. Another part (y) of the instrument is radiating to space or to the thermally indistinguishable unilluminated portion of the moon. Let us call the radiation received from space S, noting that its sign is always negative. There is also the radiation (A) of the intervening atmospheric medium, which, in theory, may be either positive or negative, though almost always it is the latter, but this is the same on all parts of the instrument whether covered by the image of the bright moon or by that of the dark sky. (The complete mathematical expression of these quantities, as affected by atmospheric absorption, is not attempted here). The final effect on such a heat-measuring instrument is the quantity

$$xM + y(-S) + (x + y)(\pm A).$$

The first term in this expression is the one we wish to know. The second term can be entirely eliminated by reducing the size of the heat-measuring instrument until it is of the exact shape and area of the lunar image, in which case $y = 0$, or by making the heat-receiving area still smaller, when by a summation of measurements in different parts of the bright image, their total (ΣM) can be readily calculated, which has been done in the present case.

The third term increases with the amount of atmospheric haze, and is never entirely absent. In order to reduce it to a minimum, I have chosen, as has already been said, only fine clear nights for the work, with a single exception which is noted in due course, where measures were made on a hazy night merely for the sake of illustrating the influence of this factor, the results not being included with the others.

The number of nights of observation could have been easily doubled or trebled by including all, good, bad, or indifferent, on which measures might have been taken. I have refrained from such a course owing to my conviction that a small number of really good observations is worth more than many times that number of measurements made under adverse circumstances. This judicious selection of opportunities is a necessary one in all measures which involve the effect of an immensely fluctuating atmospheric absorption.

Ericsson, having made many measures with his solar actinometer, rejected them all in favor of the results of a single perfect day and only published the latter. Crova, also, in his works on solar radiation, has pointed out that not the mean results, but the highest values of the solar constant, obtained under the most favorable conditions, will be the nearest to the truth, and that even these must fall short of the actuality.

The measurements of lunar radiation obtained in the following research may be depended upon, and this to a much greater degree than if a larger series of poor observations had been included, with discrepancies which would have obscured the value of the really good ones. Where all measurements are equally trustworthy, the increase in their number is a gain, but if otherwise, as in the present instance, it may be a loss. It must be remembered that the discrepancies produced by atmospheric changes far exceed those which are attributable to instrumental unreliability in this work.

The times are those of the 75th meridian W. of Greenwich, this being the adopted standard of our city

JANUARY 12, 1889.

Astronomical data.

	Eastern Time	Greenwich Time	Moon's R. A.	Moon's Dec.	Moon's H. A.	Moon's Geoc. Z. D.	Moon's Z. D. Cor. for Parallax
Beginning of Observations	8h. 53m.	13h. 53m.	4h. 25m.0	+17°56′	—21m.	22°59′	
Middle of Observations	9h. 26m.	14h. 26m.	4h. 26m.1	+18°00′	+11m.	22°36′	Average
End of Observations	9h. 58m.	14h. 58m.	4h. 27m.2	+18°03′	+42m.	24°10′	23°24′
At transit	9h. 15m.	14h. 15m.		+17°59′	0m.	22°29′	

Mean relative air-mass = 1.09
Mean barometer = 737 mm.
True air-mass = 1.057 atmosphere.

Horizontal parallax (Equatorial)	= 54′ 04″.7
„ „ (Local)	= 54′ 09″.3
Geocentric somidiameter of Moon	= 14′ 46″
Average apparent „ „ „	= 14′ 59″ = 899″
Diameter of lunar image	= 27.2 mm.
Ratio of bolometer aperture to image	= 1 : 5.04 = 0.199
Libration in lunar latitude	= 4° 41′ north
Libration in lunar longitude	= 52′ west.

i. e. mean center is N. 4°.7, E. 0°.9 from center of lunar disc.

Terminator — 44°.0 lunar west longitude (sunrise).
Moon's age = 11.2 days.

Meteorological and Instrumental Data.

Wet bulb (2 meters above the ground)	—2°.8 C. at 12h.
Dry „	— 1.5 „ „ „
Dew Point	— 7.7 „ „ „
Temperature of apartment	+10.8 „
„ „ screen	+ 6.6 „ „ 8h. 30m.
„ „ „	+ 8.8 „ „ 10h. 30m.

State of sky, from 8h. 30m. and onward, clear.
Factor reducing to standard instruments 0.982
Adopted sky-deflection (uncorrected) —22. divisions
Effective temperature of sky —54° C.

Table of Measurements.

No.	Time h. m.	Deflection Moon-screen	Moon-sky	Reduced to Standard	Approx. Air Mass	Reduced to Zenith	No.
1	8 53	64	86	84.5	1.09	85.7	1
2	8 56	38	60	58.9		59.7	2
3	8 58	5.5	27.5	27.0		27.4	3
4	9 02	35	57	56.0		56.8	4
5	05	37	59	57.9		58.7	5
6	07	—21.5					6
7	11	10.5	32.5	31.9		32.3	7
8	13	12	34	33.4		33.9	8

No.	Time		Deflection		Reduced to	Approx.	Reduced to	No.
	h.	m.	Moon-screen	Moon-sky	Standard	Air Mass	Zenith	
9	9	18	53	75	73.7		74.7	9
10		22	43	65	63.8		64.7	10
11		26	56	78	76.6		77.7	11
12		29	61	83	81.5		82.6	12
13		33	55	77	75.6		76.7	13
14		38	51	73	71.7		72.7	14
15		42	65	87	85.4		86.6	15
16		44	61	83	81.5		82.6	16
17		47	58	80	78.6		78.7	17
18		49	60	82	80.5		81.6	18
19		53	33	55	54.0		54.8	19
20		55	33	55	54.0		54.8	20
21		58	15	37	36.3		36.8	21
22	10	00	—22.5					22

1. Equator tangent to W. limb, 53° W. of center of disc.
2. Equator at central meridian.
3. Equator tangent to terminator, 28° E. of center of disc.
4. About 53° N. of center, tangent to North pole.
5. About 53° S. of center, tangent to South pole.
6. Sky near moon.
7. 35° N., tangent to terminator.
8. 35° S., tangent to terminator.
9. 35° N., tangent to N.W. limb.
10. 35° S., tangent to S.W. limb.
11. Equator at S. end of *Mare Tranquilitatis*, 25° W. of mean center.
12. Equator, including parts of *Maria F., Tr., Nect.*, 37° W. of mean center.
13. Near equator, bright region 15° W. of center.
14. Equator, bright region 10° W. of center.
15. Equator, tangent to W. limb, 53° W.
16. *Mare Tranquilitatis*, 10° N., 30° W. of mean central meridian.
17. *Mare Nectaris*, 15° S. 30° W. of mean central meridian.
18. *Mare Nectaris*, 10° S. 30° W. of mean central meridian.
19. 53° S. of center, tangent to South pole.
20. 53° N. of center, tangent to N. pole.
21. Equator, nearly tangent to terminator, 20° E. of center of disc.
22. Sky.

Positions measured more than once.

			Mean
Equator tangent to W. limb	85.7	86.6	86.2
Central meridian tangent to N. pole	56.8	54.8	55.8
Central meridian tangent to S. pole	58.7	54.8	56.8

The observations are plotted in a series of „isothermal" curves, i. e., curves which pass through regions of like radiating power, as determined by our heat measuring instrument, in fig. 7.

At 11h., the sky deflection was compared with that from screens at two different temperatures, viz.,

Cold screen at + 9°.7 C.; Sky — screen	—24	—25	—24.5
Hot „ „ + 52″.0 „ „ „	—48	—49	—48.5

Calling radiation at zero Centigrade unity, we have, according to the observations of Dulong and Petit: [1])

	Radiation at	+ 150° C.	3.01
	„ „	+ 100	2.08
	„ „	+ 50	1.44
	„ „	0	1.00
and by extension	„ „	— 50	0.68
	„ „	— 100	0.40

from which the effective temperature of the sky indicated by the above negative deflections must have been about —54° C.

JANUARY 17, 1889.

Astronomical data.

	Eastern Time	Greenwich Time	Moon's R. A.	Moon's Dec.	Moon's H. A.	Moon's Geoc. Z. D.	Moon's Z. D. Cor. for Parallax
Beginning of Observations	8h.37m.	13h.37m.	8h.42m.1	+ 19°43′	—4h.34m.	61°15′	62°04′
Middle of Observations	10h.26m.	15h.26m.	8h.46m.1	+19°36′	—2h.49m.	41°28′	42°05′
End of Observations	12h.15m.	17h.15m.	8h.50m.1	+19°28′	—1h.04m.	25°04′	25°28′
At transit	13h.21m.	(observations ended before transit).					

1) Ferrel in his „Law of Thermal Radiation" (American Journal of Science, July, 1889, p. 29), considers that the law connecting radiation and temperature, in the case of a blackened surface, is intermediate between the formula of Stephan, and that of Dulong and Petit. The question is not yet settled by thoroughly reliable experiments, and in view of the hitherto insuperable difficulties of verification at low temperatures, the above extrapolation is as accurate a one, as the present state of science will warrant.

Relative air-mass	2.134	1.347	1.107
Barometer	728mm.	729mm.	730mm.
True air-mass	2.044	1.292	1.063 atmosphere.

Horizontal parallax (Equatorial) = 55′ 32″.8
      „   „   (Local) = 55′ 37″.5
Geocentric semidiameter of Moon = 15′ 10″
Average apparent   „   „   „ = 15′ 21″ = 921″
Diameter of lunar image = 28.0 mm.
Ratio of bolometer aperture to image = 1 : 5.19 = 0.193
Libration in lunar latitude = 2° 02′ south
Libration in lunar longitude = 4° 43′ east
       i. e. mean center is S. 2°.0, W. 4°.7 from center of lunar disc.
Terminator   — 74.8 lunar east longitude (Sunset).
Moon's age   = 16.2 days.

Meteorological and Instrumental data.

Wet bulb (2 meters above the ground)   + 1°.7 C. at 7h. 30m.
Dry   „   + 3°.8   „   „   „
Dew Point   — 1°.6   „   „   „
Wet bulb (2 meters above the ground)   — 0°.4   „   „ 1h. 00m.
Dry   „   + 1°.0   „   „   „
Dew Point   — 2°.8   „   „   „
Temperature of apartment   + 13°.0   „   „ 12h. 00m.
    „   „   screen   + 11°.3   „   „   8h. 30m.
    „   „   „   + 12°.2   „ „ 12h. 20m.
State of sky, clear
Factor reducing to standard instruments   0.969
Adopted sky-deflection (uncorrected)   — 24.0   divisions
Effective temperature of sky   — 40°.0 C.

Table of Measurements.

No.	Time	Deflection Moon-screen	Moon-sky	Reduced to Standard	Approx. Air Mass	Reduced to Zenith	No.
	h. m.						
1	8 37	57	81	78.4	2.04	98.9	1
2	47	30	54	52.3	1.91	64.0	2
3	55	56	80	77.5	1.82	93.0	3
4	9 00	54	78	75.6	1.78	90.0	4

No.	Time	Deflection		Reduced to Standard	Approx. Air Mass	Reduced to Zenith	No.
		Moon-screen	Moon-sky				
	h. m.						
5	9 36	48	72	69.8	1.50	78.0	5
6	42	55	79	76.6	1.46	84.8	6
7	45	65	89	86.2	1.45	95.3	7
8	49	35	59	57.2	1.43	62.9	8
9	52	60	84	81.4	1.42	89.3	9
10	55	67	91	88.2	1.40	96.3	10
11	10 00	61	85	82.4	1.38	89.7	11
12	10 05	55	79	76.6	1.36	83.0	12
13	10	54	78	75.6	1.34	81.5	13
14	14	43	67	64.9	1.33	69.8	14
15	18	47	71	68.8	1.31	73.8	15
16	21	60	84	81.4	1.30	87.1	16
17	24	59	83	80.4	1.30	86.0	17
18	29	—21					18
19	34	60	84	81.4	1.27	86.5	19
20	37	71	95	92.1	1.26	97.7	20
21	40	68	92	89.1	1.25	94.3	21
22	44	59	83	80.4	1.24	84.9	22
23	47	47	71	68.8	1.24	72.7	23
24	50	50	74	71.7	1.23	75.6	24
25	54	50	74	71.7	1.22	75.3	25
26	57	63	87	84.3	1.21	88.3	26
27	11 02	67	91	88.2	1.20	92.3	27
28	16	63	87	84.3	1.17	87.6	28
29	20	72	96	93.0	1.16	96.3	29
30	24	74	98	95.0	1.15	98.2	30
31	30	73	97	94.0	1.14	97.0	31
32	33	73	97	94.0	1.14	97.0	32
33	36	66	90	87.2	1.13	89.8	33
34	43	42	66	64.0	1.12	65.8	34
35	46	59	83	80.4	1.11	82.5	35
36	49	59	83	80.4	1.10	82.2	36
37	12 12	57	81	78.5	1.06	79.6	37
38	15	50	74	71.7	1.06	72.7	38
39	18	— 28					39

Position.

1. Center.
2. Equator, tangent to terminator, 50° W.
3. Equator, tangent to E. limb, 53° E. of center of disc.
4. 53° S. on central meridian, tangent to S. pole.
5. 53° N. on central meridian, tangent to N. pole.
6. 53° S. on central meridian, tangent to S. pole.
7. Center.
8. Equator, tangent to terminator, 50° W. of center of disc.
9. Equator, S. of *Mare Tranquilitatis*, 27° W. of mean center.
10. Equator, S. of *Copernicus*, 20° E. of mean center.
11. Equator, S. of *Kepler*, 38° E. of mean center.
12. Equator, E. limb, 53° E. of center of disc.
13. 34° N., tangent to N.E. limb.
14. 34° N., tangent to N.W. limb.
15. 34° S., tangent to S. W. limb.
16. 34° S., tangent to S.E. limb (trifle too far W.)
17. 34° S., tangent to S.E. limb.
18. Sky near moon.
19. 53° S. on central meridian, tangent to S. pole.
20. 30° S. on central meridian.
21. Center.
22. 30° N. on central meridian.
23. 53° N. on central meridian, tangent to N. pole.
24. *Mare Tranquilitatis* (about 15° N. 30° W.)
25. „ „ „ 10° N. „
26. *Copernicus*, about 10° N., 20° E.
27. *Mare Nubium*, about 15° S., 18° E.
28. *Mare Imbrium*, about 30° S., 20° E.
29. Bright region, 6° S., 13° W.
30. „ „ 15° S., 5° W.
31. Center.
32. Equator, S. of *Kepler*, 38° E.
33. Equator, tangent to E. limb, 53° E.
34. Equator, tangent to terminator, 50° W.
35. Equator, S. of *Mare Tranquilitatis*.
36. 53° N. on central meridian, tangent to N. pole.
37. 53° S. on central meridian, tangent to S. pole.
38. N. on central meridian, tangent to N. pole.
39. Sky near moon.

Positions measured more than once.

					Mean
53° S., tangent to S. pole	90.0	84.8	86.5	79.6	85.2
53° N., tangent to N. pole	78.0	72.7	82.2	72.7	76.4
Center	98.9	95.3	94.3	97.0	96.4
Equator, 53° E., tangent to E. limb.	93.0	83.0	89.8		88.6
Equator, 50° W., tangent to terminator	64.0	62.9	65.8		64.2
Equator, S. of *Mare Tranquilitatis*	89.3	82.5			85.9
Equator, S. of *Kepler*	89.7	97.0			93.3
Mare Tranquilitatis	75.6	75.3			75.5

An average sky deflection of —24 divisions relatively to a screen at + 11°.8 C., and one of —46 divisions at 12h. 30m. by comparison with a screen at + 47°.5 C. were obtained, indicating an effective sky temperature not far from —40° C.

The isothermal curves (fig. 8), show a considerable excess of radiation from the southern or brighter hemisphere of the moon, and especially from the bright region lying west of the *Mare Nubium*, with an extension of the same along the equator towards the east to the point over which the sun is shining vertically. It may also be observed that this hotter region is in the rays of an afternoon sun; but its eccentric form and southern latitude on a globe, where seasons are almost unknown, suggests that it is due, in part at least, to some peculiarity of the surface in this region. Another fact, however, which is deducible from the isotherms of this and other dates, proves that there is a very appreciable diurnal inequality in the distribution of temperature, namely the greater steepness of the temperature gradient on the afternoon side. (Compare the morning gradients in figures 7, 10 and 12, with the afternoon gradients of figures 8 and 9). Other facts also may be adduced, which prove that there is an appreciable retention of heat by the materials of which the moon's surface is composed.

Although less than a day has elapsed since full moon, the east limb is more than 20 per cent warmer than the region along the western terminator.

JANUARY 19, 1889.

Meteorological Conditions.

Wet bulb (2 meters above the ground	— 4°.9 C. at 11h.
Dry bulb	— 3°.8 „ „ „
Dew-point	—10°.6 „ „ „
Barometer	739 mm.
Temperature of apartment	+ 9°.7 C. at 11h.
„ „ screen	+ 8°.4 „

Sky at 10h., halo of 22° to 23° radius and diffuse coronae around the moon. Uniform cirrus haze, rather thick, as indicated by the brightness of the halo. The observations have small value as lunar measures, but may be of some service by showing the effect of a rather dense but uniform haze of ice-crystals.

The moon transits at 14h. 39m. Eastern Time, and its zenith distance is over 60° at the close.

Moon's age = 19.8 days.

Table of Measurements.

Time	Moon-screen	Moon-sky	Position
h. m.			
10 04	— 7.0	+8.0	Tangent to E. limb
09	—11.5	+3.5	Tangent to W. terminator
12	—11.0	+4.0	Center
15	—11.5	+3.5	Tangent to N. pole
17	—12.5	+2.5	Tangent to S. pole.

At 10h. 26m., a mean of five comparisons of the cold screen at +8°.4 C. and the sky gave for the latter a negative deflection of —15.0, and at 10h. 40m., a mean of two comparisons with a hotter screen at +34° C. gave —33.5, showing the effective temperature of the sky to be about —15° C. according to the method already explained.

The negative deflections observed on this night are evidently the result of the radiation of the bolometer to a partially transparent screen of ice-crystals, whose temperature is not very far below that prevailing near the ground. This crystalline screen cuts off all but about 12 per cent. of the customary lunar radiation; but the characteristic distribution of heat in the lunar image (viz. bright limb about twice as hot as center, terminator cooler) is nevertheless shown in spite of the minuteness of the lunar excess over the sky deflections.

We shall have other examples of negative lunar deflections near the terminator (negative that is to say by comparison with a screen a few degrees above the freezing point, though positive by reference to the sky) which are not to be explained in this way, but which bear witness to the actual low temperature of the moon at sunrise and sunset.

JANUARY 23, 1889.

Astronomical data.

	Eastern Time h. m.	Greenwich Time h. m.	Moon's R. A. h. m.	Moon's Dec.	Moon's H. A. h. m.	Moon's Geoc. Z. D.	Moon's Z. D. Cor. for Parallax
Beginning of Observations	15 46	20 46	14 01.8	—6° 51'.6	—2 20	57° 23'	58° 13'
Middle of Observations	17 20	22 20	14 05.2	—7° 11'.0	— 50	49° 01'	49° 46'
End of Observations	18 55	23 55	14 08.7	—7° 30'.2	+ 42	48° 57'	48° 33'
At transit	18 11	23 11		—7° 21'.5	00	47° 49'	49° 42'

	15h. 46m.	17h. 20m.	18h. 11m.	18h. 55m.
Relative air-mass	1.899	1.548	1.510	1.545
Barometer	741mm.			
True air-mass	1.852	1.509	1.472	1.596

Horizontal parallax (equatorial)	= 58' 41".4
„ „ (local)	= 58' 46".4
Geocentric semidiameter of moon	= 16' 01"
Average apparent „ „ „	= 16' 12" = 972"
Diameter of lunar image	= 29.6 mm.
Ratio of bolometer aperture to image	= 1 : 5.48 = 0.182
Libration in lunar latitude	= 6° 38' south
Libration in lunar longitude	= 4° 28 east.

i. e. mean center is S. 6°.6, W. 4°.5 from center of lunar disc.

Terminator   = —0°9 lunar west longitude (sunset).

Moon's age   = 22.5 days.

Meteorological and Instrumental data.

Wet bulb (2 meters above the ground)	— 6°.2 C. at 19h. 30m.
Dry „	— 5°.7 „ „ „
Dew point	—10°.4 „ „ „
Temperature of apartment	+ 9°.0 „ „ „
„ „ screen	+ 8°.0

State of sky, clear, except a slight uniform haze, partly cirrus and partly smoke.

Factor reducing to standard instruments 1.17
Adopted sky-deflection (uncorrected)   −22 divisions
Effective temperature of sky   −42° C.

Table of Measurements.

No.	Time h. m.	Deflections Moon-Screen	Moon-Sky	Reduced to Standard	Approx. Air Mass	Reduced to Zenith	No.
1	15 46	− 9	13	15.2	1.85	18.4	1
2	50	− 1	21	24.6	1.83	29.6	2
3	53	− 5	17	19.9	1.81	23.9	3
4	56	+35	57	66.7	1.80	79.8	4
5	16 03	+23.5	45.5	53.2	1.76	63.1	5
6	10	+35	50.8	59.4	1.72	69.9	6
7		+22.5					7
8	22	+22	44	51.5	1.67	59.8	8
9	25	+27.5	49.5	57.9	1.66	67.1	9
10	28	−22					10
11	31	+32.5	54.5	63.8	1.64	73.6	11
12	36	+12	34	39.8	1.62	45.7	12
13	40	+11	33	38.6	1.61	44.2	13
14	43	− 7.5	14.5	17.0	1.60	19.4	14
15	45	− 6	16	18.7	1.59	21.3	15
16	57	− 2.5	19.5	22.8	1.56	25.8	16
17	17 00	+ 3	25	29.3	1.55	33.1	17
18	03	+ 1	23	26.9	1.54	30.3	18
19	05	+37.5	59.5	69.6	1.54	78.5	19
20	09	+13	35	41.0	1.53	46.1	20
21	11	−22	0			0.0	21
22	13	−22					22
23	18	+ 7.5	29.5	34.5	1.5	38.5	23
24	21	+10.5	32.5	38.0	”	42.4	24
25	25	+11	33	38.6	”	43.1	25
26	32	+26	48	56.2	”	62.8	26
27	35	+21.5	43.5	50.9	”	56.9	27
28	37	+29	51	59.7	”	66.7	28
29	42	+35.5	57.5	67.3	”	75.2	29

	No.	Time	Deflections Moon-Screen	Moon-Sky	Reduced to Standard	Approx. Air Mass	Reduced to Zenith	No.
		h. m.						
Trough Glass B	30	17 44	+ 4.5					30
	31	47	+ 4.7					31
	32	50	+ 2.2					32
	33	53	+ 2.0					33
	34	58	·+11	33	38.6	1.5	43.1	34
	35	18 01	+26.5	48.5	56.7	"	63.3	35
	36	08	+42	64	74.9	"	83.7	36
	37	11	+27.5	49.5	57.9	"	64.7	37
	38	14	− 1.1	20.9	24.5	"	27.4	38
	39	17	− 2.8	19.2	22.5	"	25.1	39
	40	20	− 6.5	15.5	18.1	"	20.2	40
	41	23	− 7.5	14.5	17.0	"	19.0	41
	42	25	−− 8.5	13.5	15.8	"	17.6	42
	43	28	−21					43

Position.

1. 50° N., 20° E., tangent to terminator.
2. 50° S., 20° E., tangent to terminator.
3. Equator, tangent to terminator, 13° E. of center.
4. 10° N., 52° E., tangent to limb.
5. 34° N., tangent to N.E. limb.
6. *Mare Humorum*, 20° S., 35° E.
7.   „   „
8. 45° S., 33° E., bright region tangent to limb.
9. 34° S., tangent to S.E. limb.
10. Sky near moon.
11. Equator, S. of *Kepler*, 38° E.
12. *Mare Imbrium*, 17° N., 20° E.
13. Equator, S. of *Copernicus*, 20° E.
14. Equator, tangent to terminator, 13° E. of center.
15. 30° S., tangent to terminator.
16. 30° N., tangent to terminator.
17. 50° N., 20° E., tangent to terminator.
18. 50° S., 20° E., tangent to terminator.
19. Equator, tangent to E. limb, 53° E.
20. *Mare Nubium*, 15° S., 18° E.

21. Dark face, equator, tangent to terminator.
22. Sky.
23. *Mare Imbrium*, 30° N., 18° E.
24.   „   „   „   20° E.
25.   „   „   „   22° E.
26. 23° N., tangent to limb.
27. 34° N., tangent to N.E. limb.
28. 34° S., tangent to S.E. limb.
29. Equator, tangent to E. limb, 53° E.
30. Equator, tangent to E. limb *with glass interposed*
31. Equator, tangent to E. limb „   „   „
32. Equator, tangent to terminator „ „   „
33. Equator, tangent to terminator „ „   „
34. Equator, S. of *Copernicus*, 20° E.
35. Equator, S. of *Kepler*, 38° E.
36. Equator, tangent to E. limb, 53° E.
37. Equator, S. of *Kepler*, 38° E.
38. 40° N., tangent to terminator.
39. 40° N., tangent to terminator.
40. 50° N., 20° E., tangent to terminator.
41. 50° S., 20° E., tangent to terminator.
42. 40° S., 20° E., tangent to terminator.
43. Sky.

Positions measured more than once.

				Mean
50° S., tangent to terminator		29.6	30.3	19.0	26.3
50° N., tangent to terminator		18.4	33.1	20.2	23.9
34° S., S. E. limb		67.1	66.7		66.9
34° N., N. E. limb		63.1	56.9		60.0
Equator, tangent to terminator		23.9	19.4		21.7
Equator, S. of *Copernicus*, 20° E.		44.2	43.1		43.7
Equator, S. of *Kepler*, 38° E.		73.6	63.3	64.7	67.2
Equator, tangent to E. limb, 53° E.	79.8	78.5	75.2	83.7	79.3

The isothermal curves representing these observations are given in fig. 9. They show that the decline of temperature must be very rapid while the sun is sinking through the last 15° of its altitude above the horizon. No heat whatever could be detected on the unilluminated face just outside of terminator.

A mean of three comparisons of sky and cold screen (temperature $+8°.0$ C.) gave a sky deflection of —21.7 divisions, while the mean of two comparisons with a screen at $+40°$ C. gave —39.8 divisions, from which an effective sky temperature of —42° C. is inferred.

A few measurements of the transmission of the lunar radiation by glass were made. The same specimen has been used on other occasions. It is a piece of polished plate glass 6.75 mm. thick, which has been found to transmit 81.5 per cent. of the solar radiation reflected from snow.

Uncorrected deflections, Moon-sky.

			Mean
Equator tangent to E. limb.	57.5	64.0	60.3
ditto, through glass	4.5	4.7	4.6

Transmission by glass "B" = 7.6 per cent.

			Mean
Equator tangent to terminator	14.5	20.3	17.4
ditto, through glass	2.2	2.0	2.1

Transmission by glass "B" = 12.1 per cent.

Both regions are mainly dark.

FEBRUARY 6, 1889.

Astronomical data.

	Eastern Time h. m.	Greenwich Time h. m.	Moon's R. A. h. m.	Moon's Dec.	Moon's H. A. h. m.	Moon's Geoc. Z. D.	Moon's Z. D. Cor. for Parallax
Beginning of Observations	7 53	12 53	2 34.0	$+ 9° 53'.0$	$+2\ 08$	41° 40'	42° 17'
Middle of Observations	8 21	13 21	2 34.9	$+ 9° 57'.7$	$+2\ 36$	46° 01'	46° 41'
End of Observations	8 49	13 49	2 35.8	$+10° 02'.5$	$+3\ 03$	50° 31'	51° 14'
At transit	5 41	(observations not commenced).					

	7h. 53m.	8h. 21m.	8h. 49m.
Relative air-mass	1.352	1.458	1.597
Barometer	730 mm.		
True air-mass	1.299	1.400	1.534

Horizontal parallax (Equatorial)	= 54′ 43″
„ „ (local)	= 54′ 47″.7
Geocentric semidiameter of Moon	= 14′ 56″
Average apparent „ „ „	= 15′ 06″ = 906″
Diameter of lunar image	= 27.6 mm.
Ratio of bolometer aperture to image	= 1 : 5.10 = 0.196
Libration in lunar latitude	= 6° 21′ north
Libration in lunar longitude	= 3° 56′ west

i. e. mean center is N. 6°.4, E. 3°.9, from center of lunar disc.

Terminator = + 12°.9 lunar east longitude (sunrise)

Moon's age = 6.6 days

Meteorological and instrumental data.

Wet bulb (2 meters above the ground)	—13°.9 C. at 9h. 30m.
Dry „	—12°.4 „ „ „
Dew point	—22°.6 „ „ „
Temperature of apartment	— 0°.0 „ „ 7h. 30m.
„ „ screen	+ 1°.0 „ „

State of sky, from 7h. 30m. to 8h. 50m. clear; cloudy after 8h. 50m.

Factor reducing to standard instruments 1.237

Adopted sky-deflection (uncorrected) —23 divisions

Effective temperature of sky —67° C.

Table of measurements.

No.	Time h. m.	Deflection Moon-screen	Moon-sky	Reduced to Standard	Approx. Air Mass	Reduced to Zenith	No.
1	7 53	+25	48	59.4	1.30	63.6	1
2	55	+16	39	48.2	1.31	51.7	2
3	59	— 4.5	18.5	22.9	1.32	24.6	3
4	8 02	— 3.5	19.5	24.1	1.33	25.9	4
5	8 11	— 7.0	16.7	20.7	1.36	22.4	5
6		— 5.5					6
7	15	—10	13	16.1	1.38	17.5	7
8	19	+11	34	42.1	1.39	45.9	8
9	22	— 8	15	18.6	1.40	20.3	9
10	25	+11	34	42.1	1.42	46.2	10
11	29	+21	44	54.4	1.43	59.8	11
12	31	—22.5					12
13	32	—23.5					13
14	44	— 5.0	18	22.3	1.50	24.9	14
15	49	+13.5	36.5	45.2	1.53	50.9	15

1. Equator, tangent to W. limb, 53° W.
2. Equator, 37° W. of mean center.
3. Equator, 26° W. of mean center, tangent to terminator.
4. 45° N., 32° W., tangent to terminator.
5. 45° S., 32° W., tangent to terminator.
6. 45° S., 32° W., tangent to terminator.
7. 30° S., 27° W., tangent to terminator.
8. 34° S., tangent to S. W. limb.
9. 30° N., 27° W., tangent to terminator.
10. 34° N., tangent to N. W. limb.
11. Equator, tangent to W. limb, 53° W.
12. Sky.
13. Sky.
14. Equator, tangent to terminator, 26° W.
15. 23° S., tangent to limb.

   At this point clouds formed quite suddenly.

Positions measured more than once.

			Mean
Equator, tangent to W. limb	63.6	59.8	61.7
Equator, tangent to terminator	24.6	24.9	24.8

   For isothermal curves, see fig. 10.

FEBRUARY 7, 1889.

Astronomical data.

	Eastern Time h. m.	Greenwich Time h. m.	Moon's R. A. h. m.	Moon's Dec.	Moon's H. A. h. m.	Moon's Geoc. Z. D.	Moon's Z. D. Cor. for Parallax
Beginning of Observations	7 52	12 52	3 20.3	+13° 43'.5	+1 25	32° 34'	33° 04'
Middle of Observations	8 19	13 19	3 21.2	+13° 47'.5	+1 51	35° 55'	36° 27'
End of Observations	8 45	13 45	3 22.0	+13° 51'.4	+2 16	39° 50	40° 25'
At transit	6 24	(observations not begun).					

	7h. 52m.	8h. 19m.	8h. 45m.
Relative air-mass	1.193	1.243	1.314
Barometer	734mm.		
True air-mass	1.152	1.201	1.269

Horizontal parallax (Equatorial)	$= 54'\ 22''.5$
„ „ (Local)	$= 54'\ 27''.1$
Geocentric semidiameter of Moon	$= 14'\ 5''$
Average apparent „ „ „	$= 15'\ 00'' = 900''$
Diameter of lunar image	$= 27.4$ mm.
Ratio of bolometer aperture to image	$= 1:5.07 = 0.197$
Libration in lunar latitude	$= 5°\ 50'$ north
Libration in lunar longitude	$= 2°\ 26'$ west

i. e. mean center is N. 5°.8, E. 2°.4 from center of lunar disc.

Terminator $= +0°.4$ lunar east longitude (sunrise).

Moon's age $=\quad 7.7$ days.

Meteorological and Instrumental data.

Wet bulb (2 meters above the ground)	$-10°.4$ C. at 10h. 00m.
Dry „	$-\ 9°.2$ „ „ „
Dew Point	$-14°.9$ „ „ „
Temperature of apartment	$+\ 3°.9$ „ „ 7h. 45m.
„ „ screen	$+\ 1°.3$ „

State of sky, clear until 8h. 45m., after which cirrus and halo.

Factor reducing to standard instruments 1.034.

Adopted sky-deflection (uncorrected) -19.5 divisions

Effective temperature of sky $-44°$ C.

Table of measurements.

No.	Time h.	m.	Deflection Moon-screen	Moon-sky	Reduced to Standard	Approx. Air Mass	Reduced to Zenith	No.
1	7	52	$+35.5$	53.0	54.8	1.2	57.3	1
2		55	$+31.5$					2
3	8	05	$-\ 2.0$	17.5	18.1	„	18.9	3
4		08	$+\ 0.5$	20.0	20.7	„	21.7	4
5		21	$+\ 2.7$	22.2	23.0	„	24.1	5
6		23	-19.5					6
7		35	$+29$	48.5	50.1	„	52.4	7
8		43	$+14$	33.5	34.6	„	36.2	8
9		45	$+28$	47.5	49.1	„	51.4	9

Positions.

1. Equator, tangent to W. limb, 53° W.
2. Equator, tangent to W. limb.
3. Equator, tangent to terminator.
4. 50° N., tangent to terminator.
5. 50° S., tangent to terminator.
6. Sky.
7. 34° S., tangent to S.W. limb.
8. 34° N., tangent to N.W. limb.
9. Equator, tangent to W. limb.

The formation of cirrus cloud with a lunar halo prevented further measures.

Points measured more than once.

			Mean
Equator, tangent to W. limb	57.3	51.4	54.4

The main purpose of the evening's work was to compare the transmission by glass for radiations from different regions. For this purpose, each pair of juxtaposed observations (with and without glass) constitutes one measurement, which need not be corrected either for instrumental error or atmospheric change, as the former affects all alike and the latter may be considered negligible in the short interval of time between the members of each pair.

	Equator West	Equator Terminator	Meridian 50° North	Meridian 50° South	34° S. S. W. limb	34° N. N. W. limb
a.	+33.5	− 2.0	+ 0.5	+ 2.7	+29	+14
b.	+53	+17.5	+20	+22.2	+48.5	+33.5
c.	+ 5.5	+ 2.5	+ 2.2	+ 3.5	+ 8	+ 6
d.	$\frac{5.5}{53.0}$	$\frac{2.5}{17.5}$	$\frac{2.2}{20.0}$	$\frac{3.5}{22.2}$	$\frac{8.0}{48.5}$	$\frac{6.0}{33.5}$
e.	10.4 %	14.3 %	11.0 %	15.8 %	16.5 %	17.9 %

a. Uncorrected deflection relatively to screen at +1°.3 C.
b. Uncorrected deflection relatively to sky without glass.
c. Uncorrected deflection through glass.
d. Transmission by glass "B".
e. Percentage transmission by glass "B".

The first three regions are relatively darker than the last three and give an average transmission of 11.9 %. The last three give 16.7 %.

For isothermal curves see fig. 11.

APRIL 10, 1889.

Astronomical data.

	Eastern Time	Greenwich Time	Moon's R. A.	Moon's Dec.	Moon's R. A.	Moon's Geoc. Z. D.	Moon's Z. D. Cor. for Parallax
Beginning of Observations	9h.57m.	14h.57m.	9h.34m.3	+17°25'.3	+1h.21m.	28°55'	29°23'
Middle of Observations	10h.26m.	15h.26m.	9h.35m.3	+17°21'.4	+1h.49m.	32°57'	33°28'
End of Observations	10h.55m.	15h.55m.	9h.36m.4	+17°17'.5	+2h.17m.	37°31'	38°06'
At transit	8h.33m.	(observations not commenced).					

	9h. 57m.	10h. 26m.	10h. 55m.
Relative air-mass	1.148	1.199	1.270
Barometer	733 mm.		
True air-mass	1.107	1.156	1.225

Horizontal parallax (Equatorial)	= 56' 10".4
„ „ (Local)	= 56' 15".1
Geocentric somidiameter of Moon	= 15' 20"
Average apparent „ „ „	= 15' 29" = 929"
Diameter of lunar image	= 28.3 mm.
Ratio of bolometer aperture to image	= 1 : 5.23 = 0.191
Libration in lunar latitude	= 3° 40' south
Libration in lunar longitude	= 6° 31' east.

i. e. mean center is S. 3°.7, W. 6°.5 from center of lunar disc.

Terminator — 35°.5 lunar west longitude (sunrise).

Moon's age = 10.6 days.

Meteorological and Instrumental Data.

Wet bulb (2 meters above the ground)	+8°.3 C. at 11h.
Dry „	+11.1 „ „ „
Dew Point	+ 5.6 „ „ „
Temperature of apartment	+14.2 „
„ „ screen	+16.1 „

State of sky, at 10h., clear overhead, smoke or mist at the surface.

Factor reducing to standard instruments	1.685
Adopted sky-deflection (uncorrected)	—12. divisions
Effective temperature of sky	—25° C.

Table of Measurements.

No.	Time h.	m.	Deflection Moon-screen	Moon-sky	Reduced to Standard	Approx. Air Mass	Reduced to Zenith	No.
1	9	57	+24	36	60.7	1.13	62.5	1
2	10	00	− 1	11	18.5	"	19.1	2
3		04	−12					3
4		07	− 1	11	18.5	"	19.1	4
5		10	+ 1.2	13.2	22.2	"	22.9	5
6		13	+ 1	13	21.9	"	22.6	6
7		16	− 1.2	10.8	18.2	"	18.7	7
8		22	+12	24	40.4	"	41.6	8
9		24	+18	30	50.6	"	52.1	9
10		27	+15	27	45.5	1.19	47.3	10
11		30	+15	27	45.5	"	47.3	11
12		34	+18	30	50.6	"	52.6	12
13		38	+14.5	26.5	44.7	"	46.5	13
14		42	+23	35	59.0	"	61.4	14
15		44	+ 4.5					15
16		47	+ 2.5					16
17		50	+ 4					17
18		53	+22	34	57.3	"	59.6	18
19		55	+10	22	37.1	"	38.6	19
20		58	−11					20

1. Equator, tangent to W. limb, 53° W. of center.
2. Equator, tangent to terminator, 22° E. of center.
3. Sky near moon.
4. 30° N. of equator, tangent to terminator, N. E.
5. 53° N. of equator, tangent to N. pole.
6. 53° S. of equator, tangent to S. pole.
7. 30° S. of equator, tangent to terminator, S. E.
8. Equator at center.
9. *Mare Tranquilitatis*, 10° N. of equator, 30° W.
10. *Mare Nectaris*, 10° S. of equator, 30° W.
11. 34° S. of equator, tangent to S. W. limb.
12. *Mare Serenitatis*, 30° N. 20° W.
13. 34° N. of equator, tangent to N.W. limb.
14. Equator, 37° W., including parts of *N. F.*, *Tr.*, *N.*

15. Equator, 37° W., *through glass.*
16. Equator at center   „   „
17. Equator, tangent to W. limb, *through glass.*
18. Equator, tangent to W. limb.
19. Equator at center.
20. Sky.

Positions measured more than once.

			Mean
Equator at center	41.6	38.6	40.1
Equator tangent to W. limb	62.5	59.6	61.1

Transmission	Center	= 2/22	= 9.1 %
	Equator, S. E. *M. Crisium*	= 4.5/35	= 12.9 %
	Equator, tangent to W. limb	= 4/34	= 11.8 %
Transmission by Glass "B",	mean		= 11.3 %

All of these regions are mixed, dark and bright, but on the whole mainly dark. For isothermal curves see fig. 12.

APRIL 13, 1889.

Astronomical data.

	Eastern Time h. m.	Greenwich Time h. m.	Moon's R. A. h. m.	Moon's Dec.	Moon's H. A. h. m.	Moon's Geoc. Z. D.	Moon's Z. D. Cor. for Parallax
Beginning of Observations	10 00	15 00	12 08.7	+4°19'.5	—0 59	38° 30'	39° 07'
Middle of Observations	11 15	16 15	12 11.4	+4°03'.0	+0 14	36° 34'	37° 10'
End of Observations	12 30	17 30	12 14.1	+3°46'.5	+1 26	41° 21'	42 00
At transit	11 01				0 00	36° 22'	36° 57'

	10h. 00m.	11h. 15m.	12h. 30m.	11h. 01m.
Relative air-mass	1.289	1.255	1.346	1.251
Barometer	734 mm.			
True air-mass	1.245	1.212	1.300	1.208

Horizontal parallax (Equatorial)	$= 58' 41''.9$
„ „ (Local)	$= 58' 46''.9$
Geocentric semidiameter of Moon	$= 16' 01''$
Average apparent diameter of Moon	$= 16' 12'' = 972''$
Diameter of lunar image	$= 29.6$ mm.
Ratio of bolometer aperture to image	$= 1 : 5.47 = 0.182$
Libration in lunar latitude	$= 6° 16'$ south
Libration in lunar longitude	$= 5° 59'$ east

i. e. mean center is S. 6°.3, W. 6°.0 from center of lunar disc.

Terminator $= -72°.6$ lunar west longitude (sunrise).

Moon's age $= 13.7$ days.

Meteorological and Instrumental data.

Wet bulb (2 meters above the ground)	$-1°.6$ C. at 11h. 20m.
Dry „	$+1°.0$ „ „ „
Dew Point	$-5°.5$ „ „ „
Temperature of apartment	$+11°.2$ „ „ 9h. 15m.
„ „ „	$+9°.2$ „ „ 12h. 30m.
„ „ screen	$+12°.0$ „

State of sky, clear

Factor reducing to standard instruments   1.852

Adopted sky-deflection (uncorrected) -18 divisions

Effective temperature of sky $-55°$ C.

Table of Measurements.

No.	Time h. m.	Deflection Moon-screen	Deflection Moon-sky	Reduced to Standard	Approx. Air Mass	Reduced to Zenith	No.
1	10 00	39.5	55.5	102.8	1.22	107.9	1
2	07	41.5	57.5	106.5	„	111.8	2
3	11	36.5	52.5	97.2	„	102.1	3
4	14	26	42	77.8	„	81.7	4
5	16	—16.5			„		5
6	21	29.5	45.5	84.3		88.4	6
7	24	39	55	101.9	„	107.0	7
8	26	40.5	56.5	104.6	„	109.8	8
9	29	40	56	103.7	„	108.9	9
10	32	39	55	101.9	„	107.0	10

No.	Time	Deflection Moon-screen	Moon-sky	Reduced to Standard	Approx. Air Mass	Reduced to Zenith	No.
	h. m.						
11	38	14.5	30.5	56.5	1.22	59.3	11
12	40	13	29	53.7	”	56.4	12
13	45	27.5	43.5	80.6	”	84.6	13
14	50	31	47	87.0	”	91.4	14
15	11 03	9					15
16	08	11					16
17	36	—16					17
18	41	26	42	77.8	1.25	82.5	18
19	45	16.5	32.5	60.2	”	63.8	19
20	51	38	54	100.0	”	106.0	20
21	57	39.5	55.5	102.8	”	109.0	21
22	59	20	36	66.7	”	70.7	22
23	12 01	38.5	54.5	100.9	”	107.0	23
24	03	6.5					24
25	06	6.0					25
26	09	41	57	105.6	”	111.9	26
27	12	42.5	58.5	108.3	”	114.8	27
28	15	11.5					28
29	18	12					29
30	21	39	55	101.9	1.30	109.0	30
31	23	43	59	109.3	”	117.0	31
32	25	44	60	111.1	”	118.9	32
33	27	43.5	59.5	110.2	”	117.9	33
34	30	40	56	103.7	”	111.0	34
35	32	—15.5					35

1. *Mare Serenitatis*, 30° N., 20° W.
2. *Mare Tranquilitatis*, about 10° N., 30° W.
3. Equator at center.
4. *Mare Imbrium*, 30° N., 20° E. of center.
5. Sky.
6. *Mare Nubium*, 15° S., 18° E. of center.
7. Bright region W. of *Nubium*, 15° S., 10° W. of center.
8. *Mare Nectaris*, 15° S., 33° W. of center.
9. Equator, 27° W. of center.
10. Equator, tangent to W. limb, 53° W.

11. 10° N. of equator, tangent to terminator, 50° E.
12. 10° S. of equator, tangent to terminator, 50° E.
13. 53° S., tangent to S. pole.
14. 53° N., tangent to N. pole.
15. 53° N., tangent to N. pole *through glass.*
16. 53° S.,   „   „ S.   „   „   „
17. Sky.
18. Equator, south of *Copernicus,* 20° E. of center.
19. 34° S., tangent to terminator, S. E.
20. 34° S., tangent to S. W. limb.
21. 34° N., tangent to N. W. limb.
22. 34° N., tangent to terminator N. E.
23. *Mare Serenitatis,* 30° N., 20° W.
24.   „   „   through glass.
25. *Mare Tranquilitatis,*   „   „
26. *Mare Tranquilitatis,* 10° N., 30° W.
27. Bright region, meridional, 10° S.
28.   „   „   „   through glass.
29. Bright region, 34° S. tangent to S. W. limb, through glass.
30. Equator, tangent to W. limb, 53° W.
31. Equator, 40° W. of center.
32. Equator, 27° W. of center.
33. Equator, 13° W. of center.
34. Equator, at center.
35. Sky.

Positions measured more than once.

			Mean
Equator tangent to W. limb	107.0	109.0	108.0
Mare Tranquilitatis	111.8	111.9	111.9
Equator, 27° W. of center	108.9	118.9	113.9
Mare Serenitatis	107.9	107.0	107.5
Equator at center	102.1	111.0	106.6

Transmission by glass "B".

			Dark	Bright
Limb at N. pole	9/47	19.1 %		
Bright limb at S. pole	11/43.5	25.3		25.3
Mare Serenitatis	6.5/54.5	11.9	11.9	
Mare Tranquilitatis	6.0/57.0	10.5	10.5	
Bright region S. of „ *Medius* "	11.5/58.5	19.7		19.7
Bright region, S. W. limb.	12/54	22.2		22.2

Mean transmissions; dark, 11.2 %; bright, 22.4 %.

For isothermal curves see fig. fig. 13.

APRIL 15, 1889.

Astronomical data.

	Eastern Time	Greenwich Time	Moon's R. A.	Moon's Dec.	Moon's H. A.	Moon's Geoc. Z. D.	Moon's Z. D. Cor. for Parallax
Beginning of	h. m.	h. m.	h. m.		h. m.		
Observations	11 37	16 37	13 58.4	—6°54'.7	—1 03	49°33'	50°19'
Middle of							
Observations	12 08	17 08	13 59.5	—7°01'.6	—0 33	48°06'	48°51'
End of							
Observations	12 38	17 38	14 00.7	—7°08'.3	—0 05	47°38'	48°23'
At transit	12 43	(observations ended).					

	11h. 37m.	12h. 08m.	12h. 38m.
Relative air-mass	1.566	1.520	1.506
Barometer	738 mm.		
True air-mass	1.521	1.476	1.462

Horizontal parallax (Equatorial)   = 60' 01".0

      „   „   (Local)   = 60' 06".1

Geocentric semidiameter of Moon   = 16' 23"

Average apparent   „   „   „   = 16' 36" = 998"

Diameter of lunar image   = 30.3 mm.

Ratio of bolometer aperture to image   = 1 : 5.61 = 0.178

Libration in lunar latitude   = 6°17' south

Libration in lunar longitude   = 3°25' east

i. e., mean center is S. 6°.3, W. 3°.4 from center of lunar disc.

Terminator   = — 82°.7 lunar east longitude (sunset).

Night of the full moon.

Moon's age   = 15.7 days.

Meteorological and Instrumental data.

Wet bulb (2 meters above the ground)   + 4°.3 C. at 1h.

Dry   „   + 7°.7 „ „ „

Dew Point   + 0°.6 „ „ „

Temperature of apartment   + 11°.7 „

Temperature of screen   + 14°.1 „

State of sky, clear

Factor reducing to standard instruments   1.857

Adopted sky-deflection (uncorrected)   —13.5 divisions

Effective temperature of sky   —45° C.

Table of Measurements.

No.	Time h. m.	Deflection Moon-screen	Moon-sky	Reduced to Standard	Approx. Air Mass	Reduced to Zenith	No.
1	11 40	41	54.5	101.2	1.50	113.3	1
2	43	30.5	44	81.7	„	91.5	2
3	46	27	40.5	75.2	„	84.2	3
4	48	—13.5					4
5	52	31.5	45	83.6	„	93.6	5
6	58	36.5	50	92.9	„	104.0	6
7	12 02	41.5	55	102.1	„	114.4	7
8	05	39.5	53	98.4	„	110.2	8
9	08	33	46.5	86.4	1.47	95.9	9
10	10	36	49.5	91.9	„	102.0	10
11	12	36	49.5	91.9	„	102.0	11
12	15	40	53.5	99.3	„	110.2	12
13	18	41	54.5	101.2	„	112.3	13
14	20	30	43.5	80.8	„	89.7	14
15	26	32	45.5	84.5	,,	93.8	15
16	29	29.5	43	79.9	,,	88.7	16
17	33	33	46.5	86.4	,,	95.9	17
18	35	32.5	46	85.4	,,	94.8	18
19	38	43	56.5	104.9	,,	116.4	19
20	41	—13.5					20

1. Equator at center.
2. Equator, tangent to E. limb, 53° E.
3. Equator, tangent to W. limb, 53° W.
4. Sky.
5. 53° N., tangent to N. pole.
6. 30° N. meridional.
7. Bright region west of *Mare Nubium*, 15°S., 10°W·
8. 30° S. meridional.
9. 53° S., tangent to S. pole.
10. *Mare Imbrium*, 30° N., 20° E.
11. *Mare Serenitatis*, 30° N., 20° W.
12. Equator, S. of *Mare Tranquilitatis*, 27° W.
13. Equator, S. of *Copernicus*, 20° E.
14. 34° N., tangent to N. E. limb.
15. 34° N., tangent to N. W. limb.
16. 34° S., tangent to S. W. limb.
17. 34° S., tangent to S. E. limb.
18. 34° S., tangent to S. E. limb.
19. Equator at center.
20. Sky.

SUMMARY OF RESULTS.

These measures having been made only 1/4 day after full moon, may be taken as a typical example of the distribution of heat in the lunar image at the full.

It will be interesting to compare the east and west limbs, by means of the following table:

S.	95.9	S. E.	95.4	N. E.	89.7	E.	91.5	Mean E. limb	92.2	Cen.	113.3
N.	93.6	S. W.	88.7	N. W.	93.8	W.	84.2	W. „	88.9	„	116.4
Mean	94.8	Mean	92.1	Mean	91.8	Mean	87.9			Mean	114.9

It will be seen that there is very little difference between the east and west limbs, the east being only a trifle hotter, although, as has been shown from the observations of Jan 17, the excess of heat at the east limb may become as large as 1/5 in less than one day after the full.

The progressive decrement of heat from higher to lower latitudes, though small, is fairly uniform, a mean latitude of 53° giving 95 divisions; 30°, 92 divisions, and 0°, 88 divisions. Since these regions of the moon are here under a sun of almost exactly the same altitude and differ mainly in the greater length of time during which the higher latitudes have been exposed to the solar rays, there is here some indication of a small heat storing action, by which an excess of heat (not greater however than 10%) is accumulated after many days of continuous sunshine.

A notable feature in the distribution of heat in the image of the full moon is its close approach to uniformity. The heat in the circumferential zone differs from that at the center by only about 20 per cent. In this respect the thermal image approaches the visual, in which, as is well known, the distribution of light is such as to suggest to the imagination a flat disc rather than a sphere. The specially brilliant rim is too narrow to have its heat tested by the bolometer, and in general the comparison of the relative heat in dark and bright areas is not easy, because the few regions suitable for the comparison are seldom under precisely the same conditions of illumination. Perhaps the best examples are the meridional areas 30° S. and 30° N. of the equator, the former mainly bright (the area around „*Regiomontanus*") and the latter mainly dark („*palus nebularum*").

Measures along a parallel but near the meridian at epochs not far from the full may also be included. Of these there is an excellent example in the comparison of the bright region W. of *Mare Nubium*, 15° S. of equator, 10° W. of meridian, and a similar dark area in the Mare Nubium itself. (15° S., 18° E.) A comparison of the bright region N. of *Schickard* (34° S.) and the dark region N. of *Herodotus* (34° N.) is less desirable on account of the near proximity of the E. limb, but has been included in the following table in which, omitting observations made close to the terminator, where a slight error in position is liable to give a considerable variation in the heat, there are given six suitable comparisons of dark and bright regions under nearly identical conditions.

Position	Date	Bright	Dark	Ratio Bright-dark
On the central meridian, latitude 30°	Jan. 17	S.	N.	1.15
	Apr. 15	97.7	84.9	
		110.2	104.9	1.06
On same parallel, 15° S. latitude	Jan. 17	10° W. of Cen.	18° E. of Cen.	
	Terminator +74°.8		Mare Nub.	
	near W. limb	98.2	92.3	1.06
	Apr. 13	107.0	88.4	1.21
	Terminator —72°.6			
	near E. limb			
Latitude 34°	Jan. 17	S.	N.	
Near 50th meridian		86.0	81.5	1.06
Tang. to E. limb	Jan. 23	66.9	60.0	1.12

$$\text{Mean, } \frac{\text{bright}}{\text{dark}} = 1.11$$

On the whole there seems to be some evidence that bright regions radiate a little more than dark, at least during the middle of the lunar day; but if the following measures along the terminator are to be trusted, the effect is reversed with a low altitude of the sun. I must, however, repeat the caution that such measures are far more liable to error than those made at a distance from the terminator.

			Bright	Dark	Ratio	
		At Terminator				
		Latitude 30°—34°				
Sunrise	Jan. 12	(34°)	33.9	32.3	1.05	
Sunset	Jan. 23	(30°)	21.3	25.8	0.83	Mixed, bright and dark
Sunrise	Feb. 6	(30°)	17.5	20.3	0.86	
Sunrise	Apr. 10	(30°)	18.7	19.1	0.98	
Sunrise	Apr. 13	(34°)	63.8	70.7	0.90	
	Mean ratio, bright—dark			=	0.92	

In the preceding comparison, no attempt is made to separate the two kinds of radiation (emitted and reflected), but we come now to evidence, obtained by measurements of glass transmission, which has a direct bearing on this point, and which is also indirectly connected with the problem of the change in the total lunar radiation at different phases.

COMPARISON BETWEEN THE PHASE-CHANGES OF EMITTED AND REFLECTED RADIATIONS.

It has been pointed out, that the relatively greater apparent angular area occupied by the colder parts of the moon, in the partial phases, will diminish the total radiation somewhat as the greater proportion of shadow in these colder parts diminishes the reflected light.

At the first quarter, for example, a belt 10° wide on the illuminated side of the terminator occupies 0.22 of the entire illuminated area visible from the earth, here a half disc, while at the full the corresponding zone is only 0.03 of the whole illuminated part, considered, according to the appearance, as a flat circular disc In like manner the shadows, not merely of hills and mountains, but of the microscopical elevations in what would ordinarily be called a smooth surface, occupy a considerable portion of the surface presented to us at the first quarter, while at the full, they disappear.

A complete discussion of the change in the total radiation with the phase, depending as it does upon a knowledge of the distribution of temperature at various parts of the moon's surface, as well as upon the purely geometrical consideration of the presentation of these regions of different temperature towards the earth, and upon the physical laws of radiation from different surfaces at every angle of inclination, need not be attempted here; but a preliminary comparison between the phase-curve given by Zöllner for the moon's light (see „Photometrische Untersuchungen," Tafel IV, und S. 198—199) and that to be deduced from these heat observations for the total radiation is desirable, and before proceeding to the final and conclusive proof of a difference in the variation of the visible and invisible components of the moon's rays, it may be well to give some evidence of another sort which leads to a similar result.

The opacity of glass for most of the rays of greater wave-length than 3μ (a particular specimen of glass was found to be completely opaque to all radiation beyond 6μ and nearly so to all beyond 3μ) enables us to roughly distinguish between the proper lunar radiations and the reflected ones; because the larger part (fully 97 %) of the sun's rays which are presented for reflection have a wave-length less than 3μ while the peculiar radiations of the moon are longer than 3μ. Consequently, any marked difference between the increment of the proper and the reflected radiations will be at once indicated by variation with the phase in the transmission of the lunar rays by glass. It is true that a very small difference between the phase-curves might be masked by a variation in the transmissibility of glass, which is liable to occur from a change in the composition of the lunar beam, due to the fluctuation of atmospheric moisture, a large amount of water-vapor in the air cutting off an undue proportion of the extreme infra-red rays to which glass is opaque, and thus increasing the glass transmission for the total radiations. Also, since, as we shall see, the bright parts of the moon are richer than the dark, in reflected rays of wave-length less than 3μ, it is necessary to restrict comparisons to regions similar in this respect.

The tabular statement of glass transmissions which follows shows very little change in the quality of radiation from dark areas with the progress of the phase, but a somewhat larger variation in the case of bright regions. The relative humidity and dew-point of the air 2 meters above the ground are given, although these tell us less than we could wish about the condition of the entire air-column as to water-vapor. Since, however, the measures were made at the season when atmospheric moisture is least, the result is probab fairly trustworthy.

Percentage Transmission by Glass "B" (6.75 mm. thick).

Date 1889	Jan. 23	Feb. 7	Apr. 10	Apr. 13
Moon's age	22.5 days, about 3rd quarter	7.7 days, abt 1st quarter	10.6 days, half-way 1st to full	13.7 days, nearly full
Transmission By Glass "B"	Dark 9.9 % Mean of two different regions	Dark 11.9% (Mean of 3) Bright 16.7 % (Mean of 3)	Mixed, rather dark 11.3 % (Mean of 3)	Dark, 11.2 % (Mean of 2) Bright 22.4 % (Mean of 3)
Relative Humidity	86 %	60 %	66 %	56 %
Dew-Point	—10°.4 C.	—14°.9 C.	+5°.6 C.	—5°.5 C.

Transmission by glass for bright regions, near the limb and near the center, differs no more than does the transmission for bright regions at different parts of the limb, as the following measures show:

Transmission for limb at S. pole	25.3 %
" " " S. W. quadrant	22.2
" " region near center, S. of *Medius*	19.7

all being bright areas. The order of these transmissions is the same in which eye estimates of the brightness of the different regions would place them, which can hardly be due to chance, since repeated measures on the same region indicate that the observations are usually trustworthy within 5 %, or one in the unit's place of the above figures.

We may conclude from these comparisons that the bright regions reflect nearly twice as much as the dark (some more and some less, according to the degree of apparent brightness), and that there is a somewhat larger proportion of reflected rays from the full moon than from the moon at the first quarter.

The diminution of heat from the moon at the time of a total eclipse has been shown by BOEDDICKER (see Trans. R. Dublin Soc. Vol. III, p. 328) to follow a different law from that of the light. „The minimum of the heat-effect falls decidedly later than the minimum of illumination, which may be supposed to coincide with the middle of the eclipse." This observation is in agreement with others which have been made here. In four concordant measurements on the night of January 16, 1889, the instrumental conditions being the same as on the next night, the umbra of the eclipsed moon was found sufficiently hot to appreciably affect a bolometer covering as little as one-thirtienth of the moon's disc, by an amount equal to about one per cent, of the heat which was to be expected from the full moon, the radiations being of a quality to which glass was impermeable. The heat in the penumbra at the edge of the umbra was at the same time hardly more than one-half that at the edge of the terminator in the uneclipsed moon.

The presence of even so much as one per cent of radiant heat in the image of the eclipsed moon indicates a difference in the behavior of the invisible heat-producing radiations of great wave-length and that of the light of the moon, since it is certain that the umbra does not give anything like one per cent, of the *light* of the full moon; but this amount is still so small that we could hardly expect any sensible change in the epoch of maximum heat to be produced by this retention of heat by the moon's surface.

A further comparison of the phase-curve for total lunar radiations and that for the luminous rays, will be given after the discussion which follows.

SUMMARY OF OBSERVATIONS BY SERIES.

The next table gives a general view of those measures which admit of being arranged in series, viz. from south to north, along either east or west limbs, terminator or central meridian, and from west to east along the equator. This arrangement presents the more salient features to the eye at a glance, although necessarily omitting some interesting measures made at irregularly scattered positions.

JAN. 12.

Along west limb	53° S.	35° S.	0°	35° N.	53° N.		
Heat	56.8	64.7	86.2	74.7	55.8		

Along terminator (E.)	53° S.	35° S.	0°	35° N.	53° N.		
Heat	56.8	33.9	27.4	32.3	55.8		

Along equator	52° W.	37° W.	25° W.	10° W.	0°	20° E.	28° E.
Heat	86.2	82.6	77.7	72.7	59.7	36.8	27.4

JAN. 17.

Along east limb	53° S.	34° S.	0°	34° N.	53° N.
Heat	85.2	86.0	88.6	81.5	76.4

Along terminator (W.)	53° S.	34° S.	0°	34° N.	53° N.
Heat	85.2	73.8	64.2	69.8	76.4

Along central meridian	53° S.	30° S.	15° S.*	0°	30° N.	53° N.	
Heat	85.2	97.7	98.2	96.4	84.9	76.4	
Along equator	50° W.	27° W.	13° W.†	0°	20° E.	38° E.	53° E.
Heat	64.2	85.9	96.3	96.4	96.3	93.3	88.6

*5° W. from meridian. †About 6° south.

JAN. 23.

Along east limb	50° S.	45° S.	34° S.	0°	23° N.	34° N.	50° N.
Heat	26.3	59.8	66.9	79.3	62.8	60.0	23.9
Along terminator (center)	50° S.	30° S.	0°	30° N.	50° N.		
Heat	26.3	21.3	21.7	25.8	23.9		
Along equator	13° E.	20° E.	38° E.	52° E.			
Heat	21.7	43.7	67.2	79.3			

FEBRUARY 6.

Along west limb	45° S.	34° S.	23° S.	0°	34° N.	45° N.
Heat	22.4	45.9	50.9	61.7	46.2	25.9
Along terminator (30° W.)	45° S.	30° S.	0°	30° N.	45° N.	
Heat	22.4	17.5	24.8	20.3	25.9	
Along equator	53° W.	37° W.	26° W.			
Heat	61.7	51.7	24.8			

FEB. 7.

Along W. limb	50° S.	34° S.	0°	34° N.	50° N.
Heat	24.1	52.4	54.4	36.2	21.7

APR. 10.

Along W. limb	53° S.	34° S.	0°	34° N.	53° N.
Heat	22.6	47.3	61.1	46.5	22.9
Along terminator (20° E.)	53° S.	30° S.	0°	30° N.	53° N.
Heat	22.6	18.7	19.1	19.1	22.9
Along equator	53° W.	37° W.	30° W.*	0°	22° E.
Heat	61.1	61.4	49.7	40.1	19.1

*Mean of 10° N. and 10° S.

APR. 13.

Along W. limb	53° S.	34° S.	0°	34° N.	53° N.
Heat	84.6	106.0	108.0	109.0	91.4

Along terminator (E.)	53° S.	34° S.	10° S.	0°*	10° N.	34° N.	53° N.
Heat	84.6	63.8	58.4	57.9	59.3	70.7	91.4

Along central meridian	53° S.	10° S.	0°	53° N.
Heat	84.6	114.8	106.6	91.4

Along equator	53° W.	40° W.	27° W.	13° W.	0°*	20° E.	50° E.
Heat	108.0	117.0	113.9	117.9	106.6	82.5	57.9

*Mean of 10° N. and 10° S.

APR. 15.

Along east limb	53° S.	34° S.	0°	34° N.	53° N.
Heat	95.9	95.4	91.5	89.7	93.6

Along west limb	53° S.	34° S.	0°	34° N.	53° N.
Heat	95.9	88.7	84.2	93.8	93.6

Along central meridian	53° S.	30° S.	0°	30° N.	53° N.
Heat	95.9	110.2	114.9	104.0	93.6

Along equator	53° W.	27° W.	0°	20° E.	53° E.
Heat	84.2	110.2	114.9	112.3	91.5

MAPPING AND SUMMATION OF HEAT-CONTOURS.

All of the observations having been plotted in the series of maps shown in figs. 7 to 14, and contour lines having been drawn through points possessing the same heat, the remainder of the process has been to divide the lunar image into small squares as is done on a plan in computing earthwork, or the volume of earth in a hill, and having estimated the height (thermal energy) of each paralelopiped from the contour line map, the sum of the volumes has been taken to represent the total heat. The next table gives this heat for the separate thermal zones into which the image has been divided in these maps.

Jan. 12

Phase-angle —45°.6

Zone			Heat
Over	85	div.	13.9
80	to	85	30.7
70	„	80	34.1
60	„	70	42.1
50	„	60	34.2
40	„	50	11.8
30	„	40	9.8
20	„	30	4.9
0	„	20	3.9
	Total		184.9

Jan. 17

Phase-angle +10°.9

Zone			Heat
Over	95	div.	35.7
90	to	95	50.1
80	„	90	104.4
70	„	80	59.4
60	„	70	16.8
0	„	60	24.5
	Total		290.9

Jan. 23

Phase-angle +87°.0

Zone			Heat
Over	80	div.	11.4
70	to	80	18.1
60	„	70	24.2
50	„	60	15.5
40	„	50	10.2
30	„	40	3.0
20	„	30	2.8
0	„	20	4.1
	Total		89.3

Febr. 6

Phase-angle —99°.0

Zone			Heat
60	to	70 div.	11.7
50	„	60	17.5
40	„	50	11.1
20	„	40	6.4
0	„	20	4.2
	Total		50.9

Febr. 7

Phase-angle —88°.0

Zone			Heat
50	to	60 div.	24.6
40	„	50	15.9
30	„	40	9.7
20	„	30	8.6
0	„	20	5.0
	Total		63.8

Apr. 10

Phase-angle —81°.0

Zone			Heat
Over	60	div.	13.7
50	to	60	31.8
40	„	50	26.8
30	„	40	17.0
20	„	30	13.6
10	„	20	5.1
0	„	10	1.4
	Total		109.4

Apr. 13

Phase-angle —23°.4

Zone			Heat
Over	115	div.	11.0
110	to	115	38.8
100	„	110	86.8
90	„	100	50.6
80	„	90	48.7
70	„	80	37.5
60	„	70	25.2
50	„	60	9.4
30	„	50	6.0
0	„	30	2.0
	Total		316.0

Phase-angle + 3°.9

Zone			Heat
Over	110	div.	77.9
100	to	110	104.1
90	„	100	75.1
80	„	90	30.9
70	„	80	17.5
50	„	70	22.0
0	„	50	11.1
	Total		338.6

From these sum-totals the ordinates for the phase-curve (fig. 15) have been taken. The abscissae are phase-angles, negative before, and positive after the full moon. By the „phase-angle" is meant the angular distance between the center of the lunar disc and the point in the heavens opposite to the sun. The maps having been drawn on a common scale, the observations are reduced to the mean distance of the moon by the process of summation.

The principal results of the preceding observations are embodied in the final table, which gives the ordinates of the smooth curve for each tenth degree of the phase-angle, first in our arbitrary units, and then as a percentage of the heat at full moon, while the last two columns show the variation of the moon's light, as observed by Zöllner, and that of the moon's heat, in the curve given by Lord Rosse (in Phil. Trans. R. Soc. 1873, Plate 48).

FINAL TABLE.

Phase-angle from Full Moon	Total Heat	Percentage of total heat at full	Light according to Zöllner	Heat according to Lord Rosse
—100°	50	14.9		11.4
— 90°	63	18.8		15.4
— 80°	75	22.3		21.9
— 70°	92	27.4	14.4	29.4
— 60°	116	34.5	22.3	37.1
— 50°	153	45.5	32.1	46.2
— 40°	200	59.5	43.7	56.4
— 30°	259	77.1	56.8	68.6
— 20°	310	92.3	70.9	83.6
— 10°	334	99.4	85.5	97.9
— 0°	336	100.0	100.0	100.0
+ 10°	315	93.8	85.5	91.4
+ 20°	282	83.9	70.9	80.5
+ 30°	246	73.2	56.8	69.3
+ 40°	211	62.8	43.7	58.4
+ 50°	180	53.6	32.1	47.9
+ 60°	151	44.9	22.3	38.3
+ 70°	126	37.5	14.4	32.3
+ 80°	104	31.0		25.5
+ 90°	84	25.0		19.8
+100°	67	19.9		13.7

This table shows conclusively, first, that visible rays form a much larger proportion of the total radiation at the full than at the partial phases, the maximum for light being much more pronounced than that for the heat.

Next, as has been foreseen from the eccentricity of the heat areas, their greater extension toward the western limb, and the greater steepness of the sunset than of the sunrise gradient, the diminution of the heat from the full to the third quarter is slower than its increase from the first quarter to the full.

Finally, there is a fair agreement between these results and those of Lord Rosse which extends even to some minor details such as the attainment of the highest heat a little before the full. This deviation of the maximum from strict symmetry is probably real, and is perhaps attributable to the greater proportion of bright areas in the western half of the moon, the brighter parts, as we have learned, giving a larger radiation under a high sun, than the dark. It is possible that this effect is reversed with a low sun, the dark parts radiating more than the bright, and that the greater heat of the lunar afternoon may be due less to a retention of heat, than to the greater darkness of the region exposed to view at that time. That there must be some retention of heat by the substances of the lunar surface, cannot, however, be doubted in view of the contrast in the heat of polar and equatorial regions under identical illumination, which has been described in connection with the observations of April 15th.

Previous investigations have dealt with the heat produced by the radiation from the entire moon, but the method pursued in the present research has been to study the thermal effect of small portions of the lunar disc, thereby eliciting many new facts concerning the distribution of heat in the moon and its variation through the lunar day for each of these circumscribed regions. The relative radiations of dark and bright surfaces under high or low sun, and of high and low latitudes, in the lunar morning, afternoon, or noon, have thus been measured, and the accompanying maps present (it is believed for the first time) a picture of the distribution of heat on a planet, where seasons and the climatic influences of land and water must be unknown.

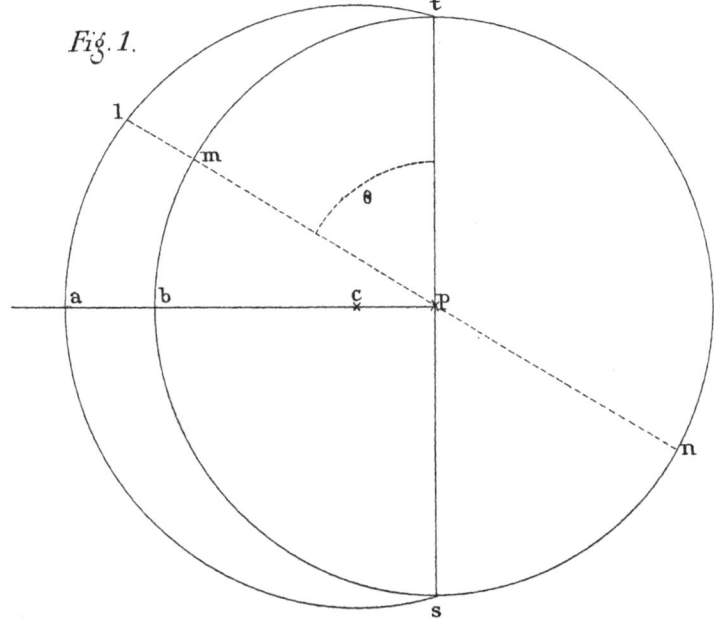

Fig. 1.

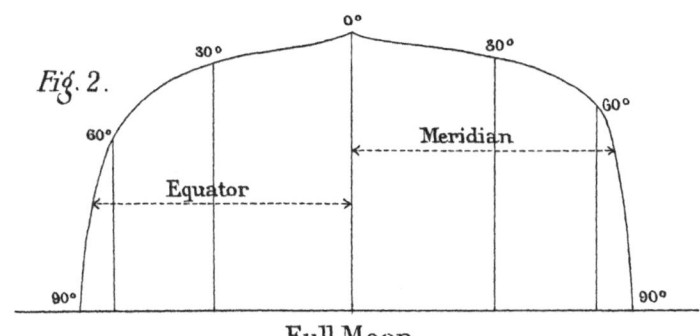

Fig. 2.

Full Moon

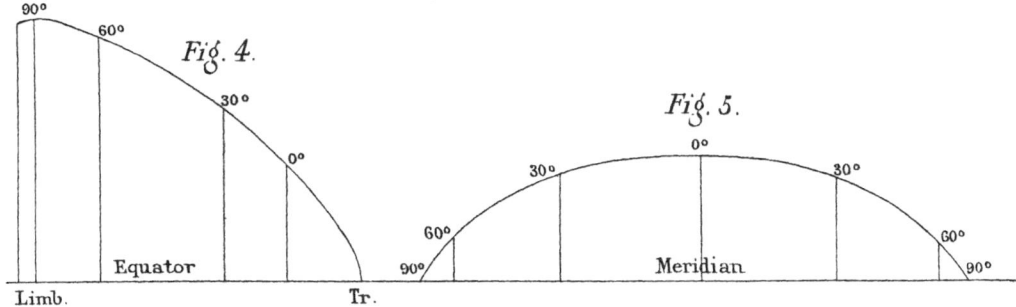

Fig. 4.

Fig. 5.

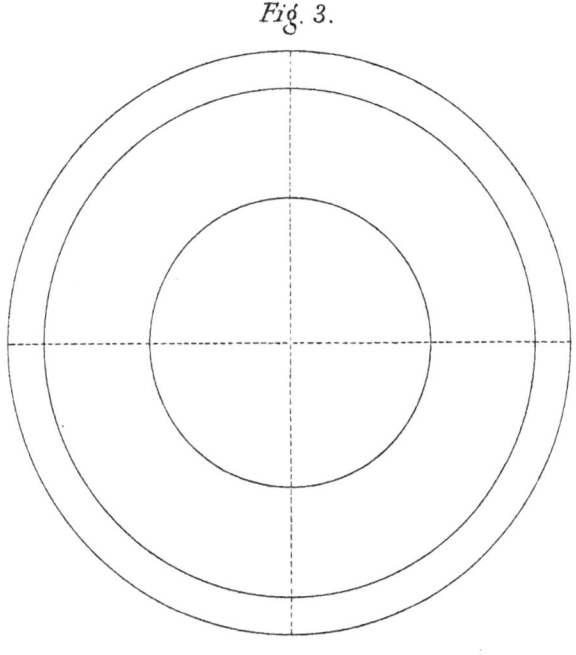

Fig. 3.

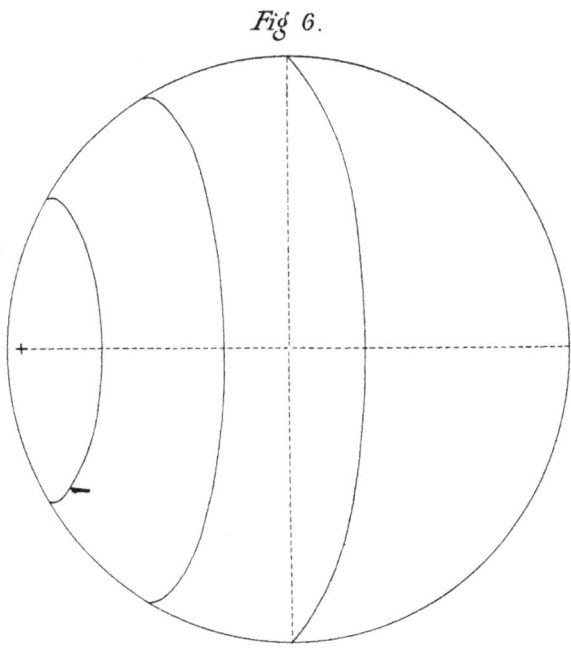

Fig 6.

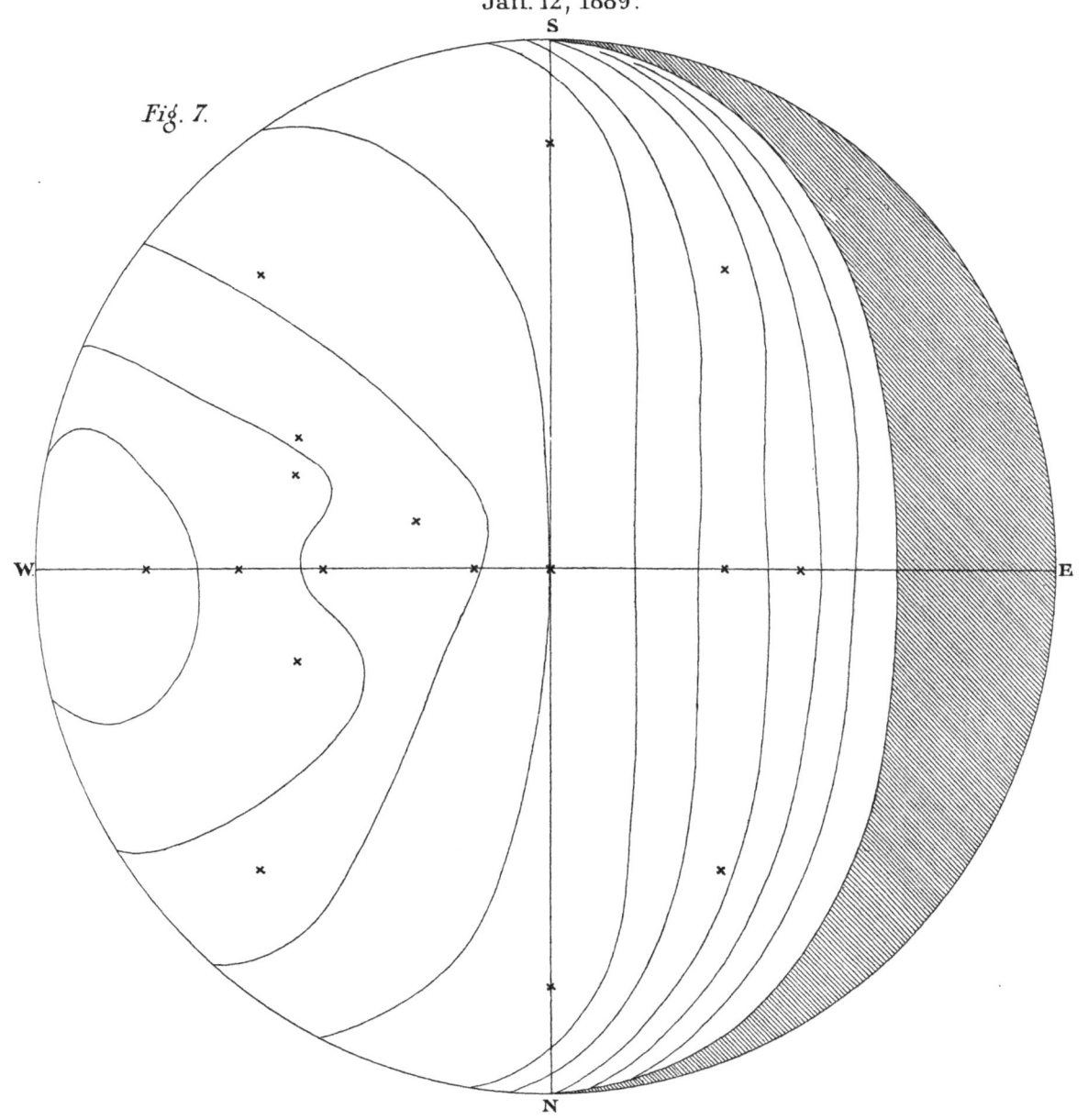

Distribution of Heat in Moon.
Jan. 12, 1889.

Fig. 7.

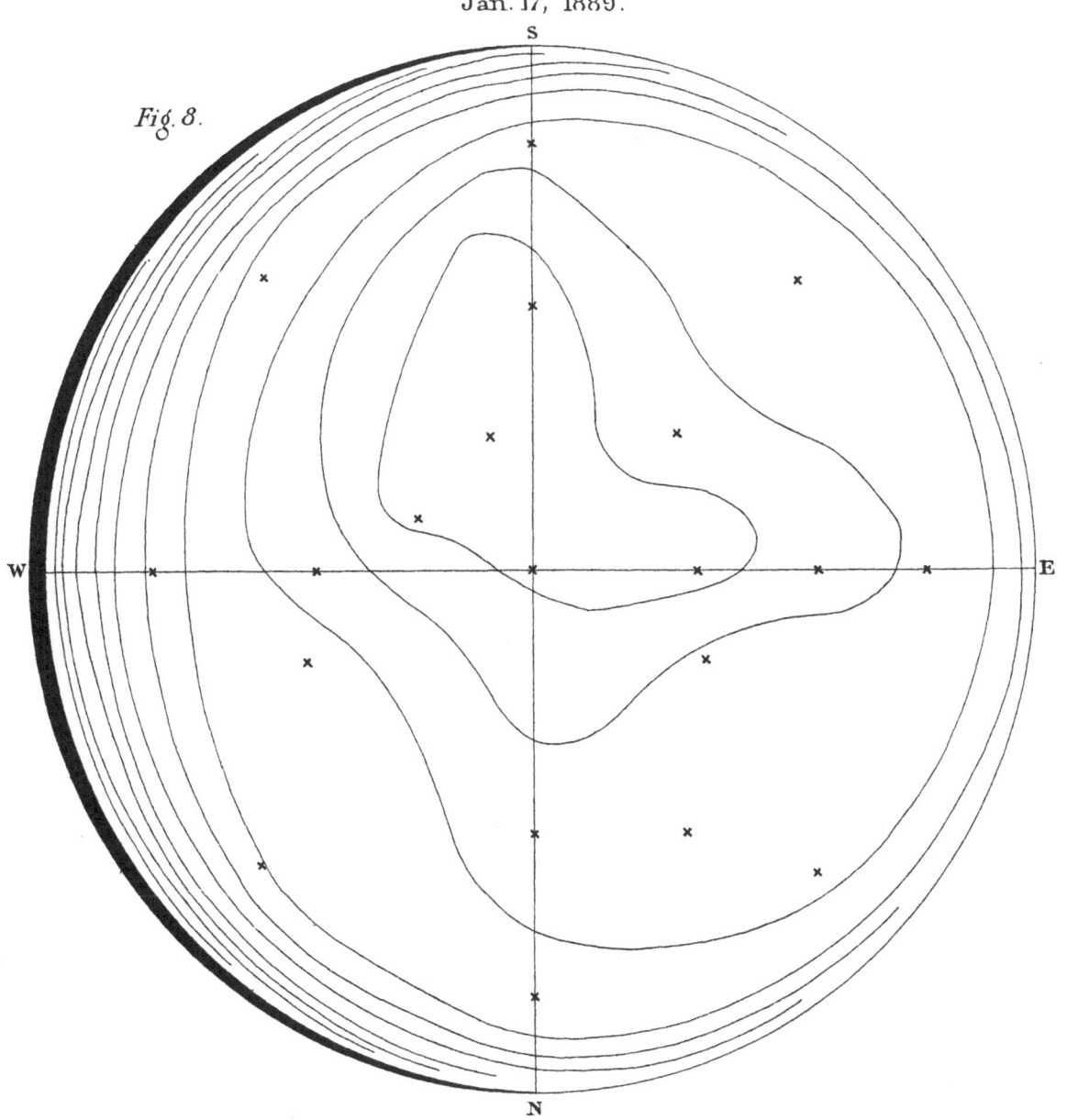

Distribution of Heat in Moon.
Jan. 17, 1889.

Fig. 8.

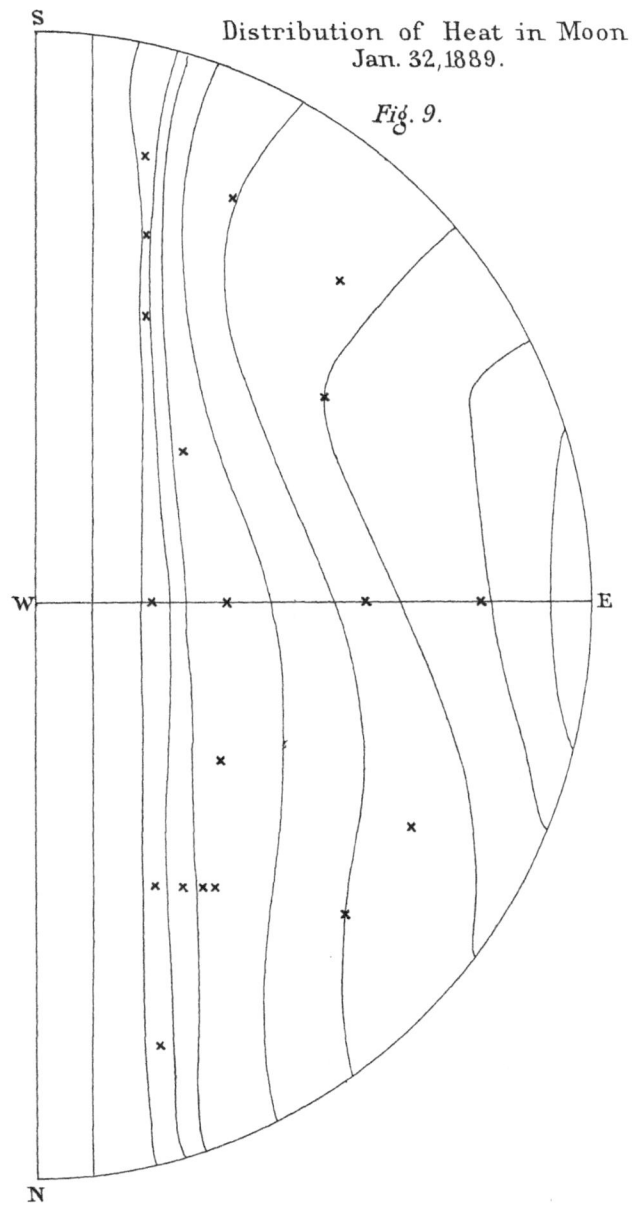

Distribution of Heat in Moon.
Jan. 32, 1889.

Fig. 9.

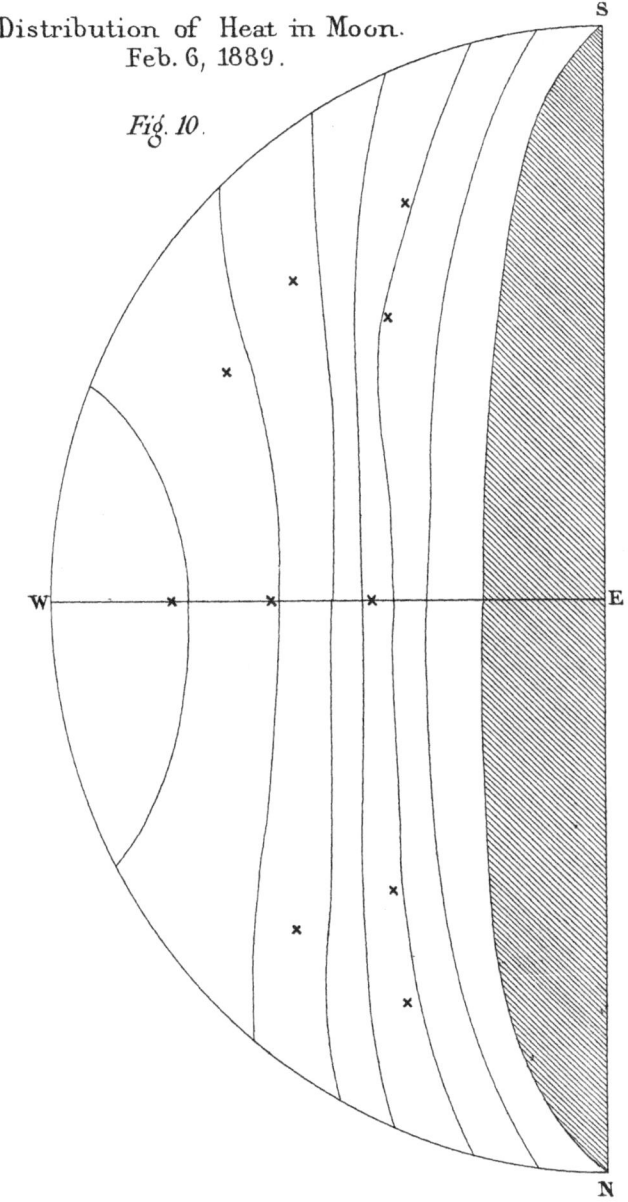

Distribution of Heat in Moon.
Feb. 6, 1889.

Fig. 10.

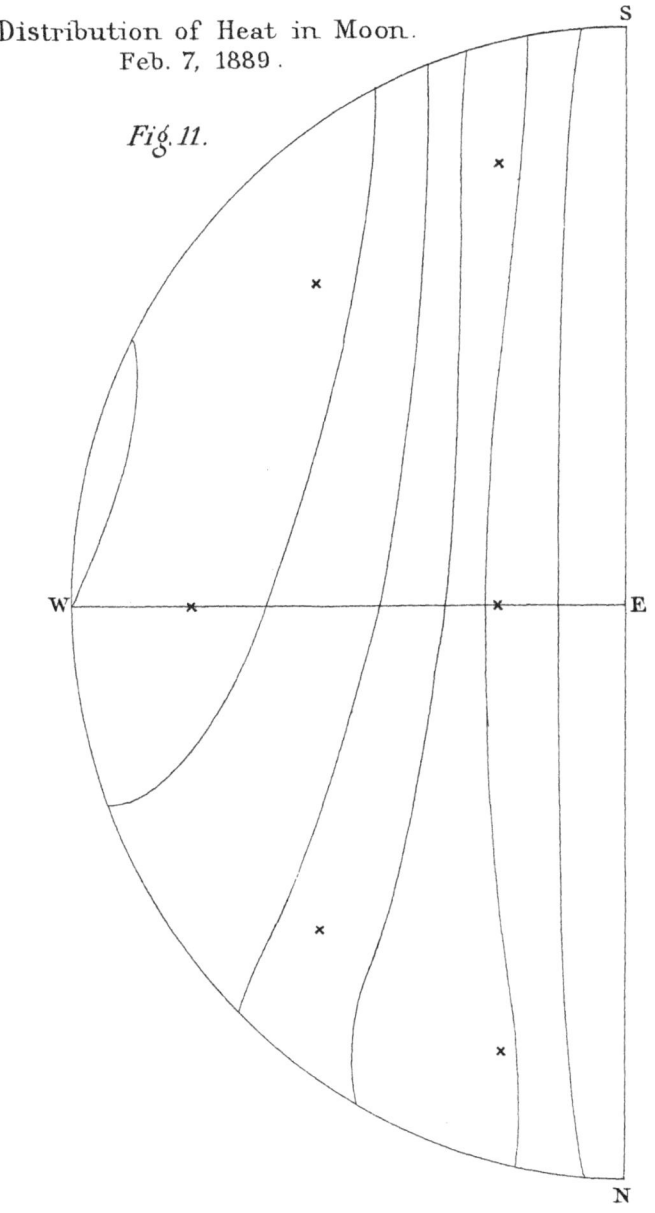

Distribution of Heat in Moon.
Feb. 7, 1889.

Fig. 11.

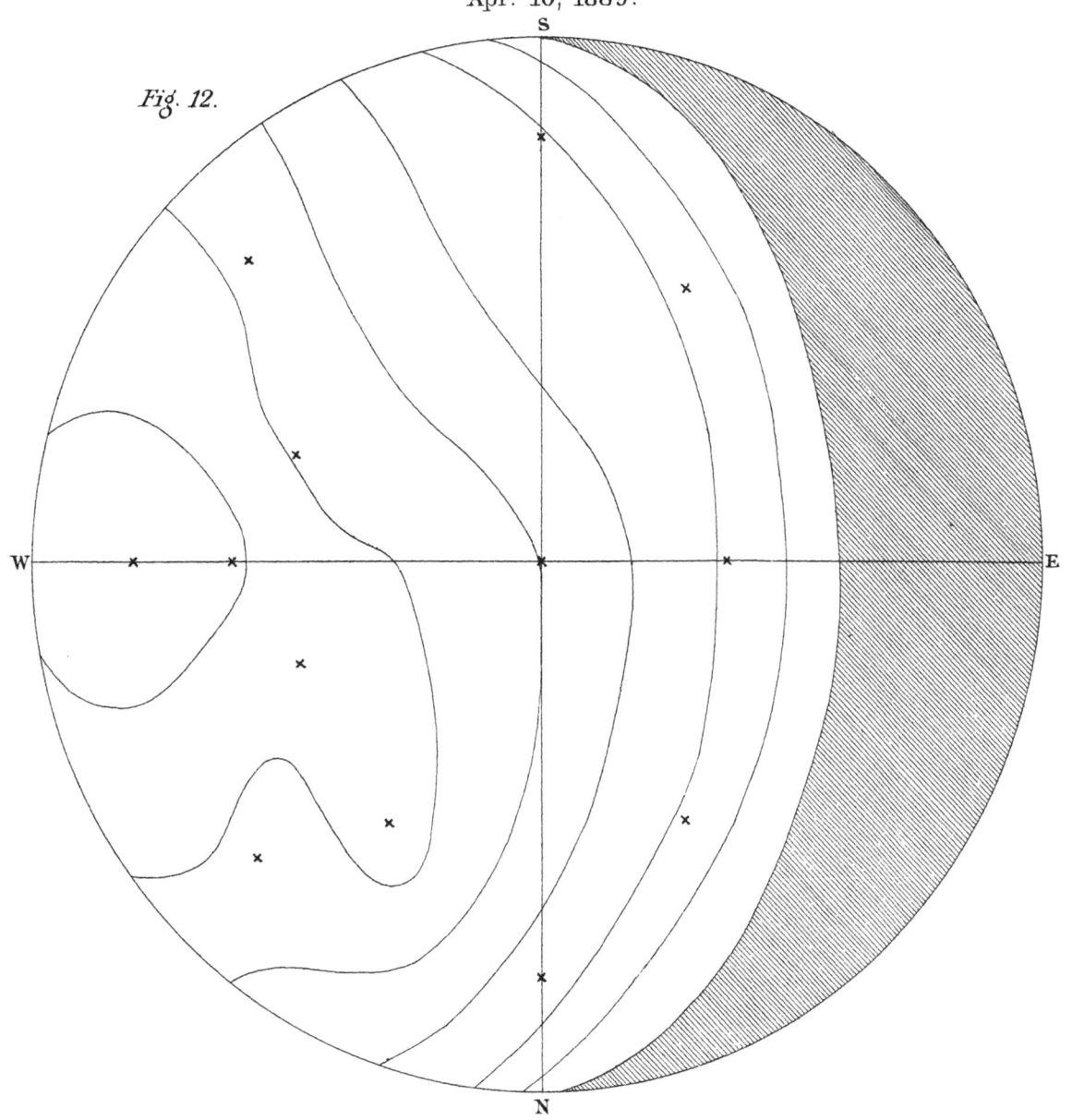

Distribution of Heat in Moon.
Apr. 10, 1889.

Fig. 12.

Distribution of Heat in Moon.
April 13, 1889.

Fig. 13.

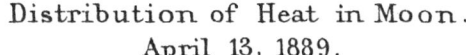

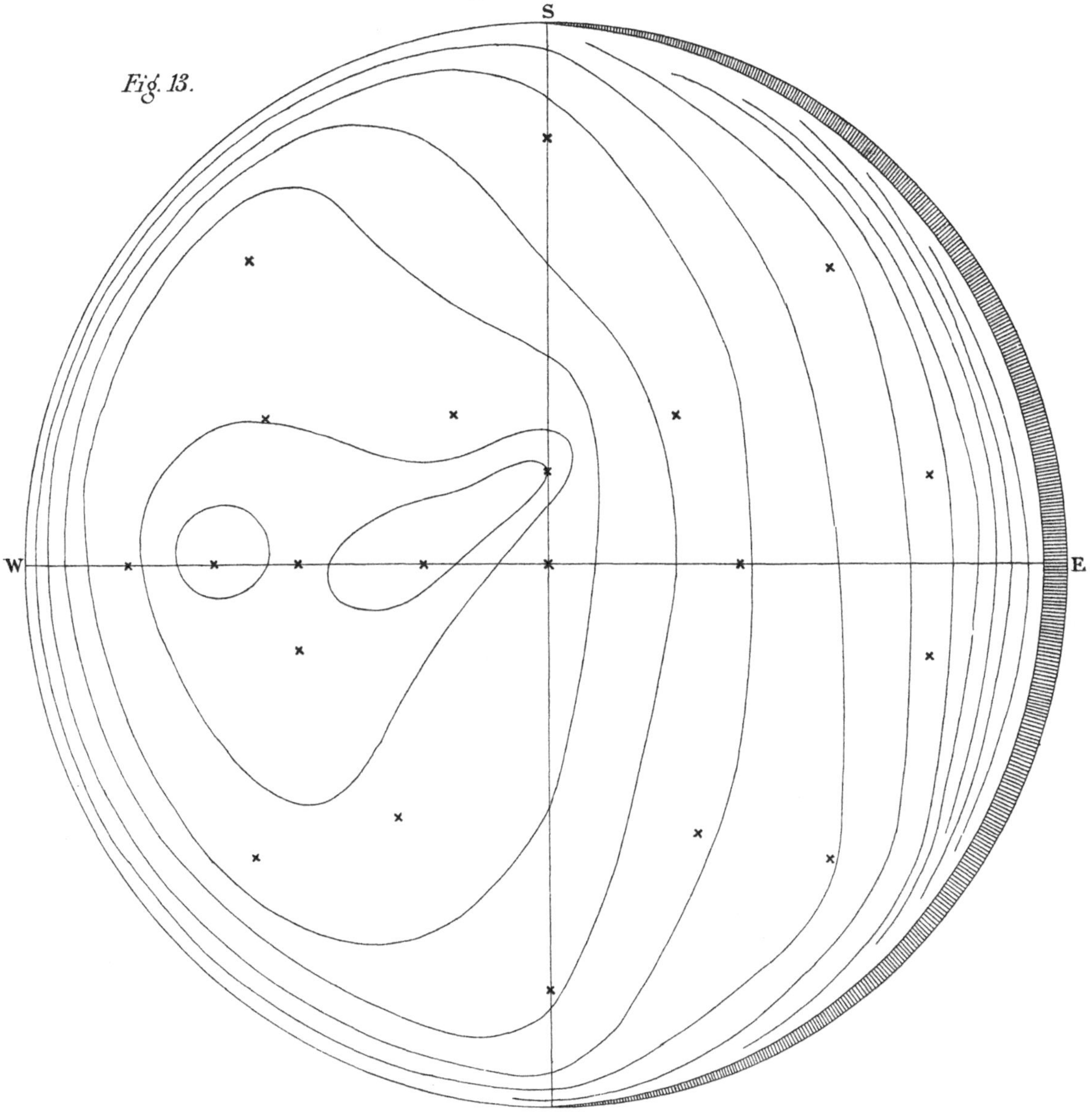

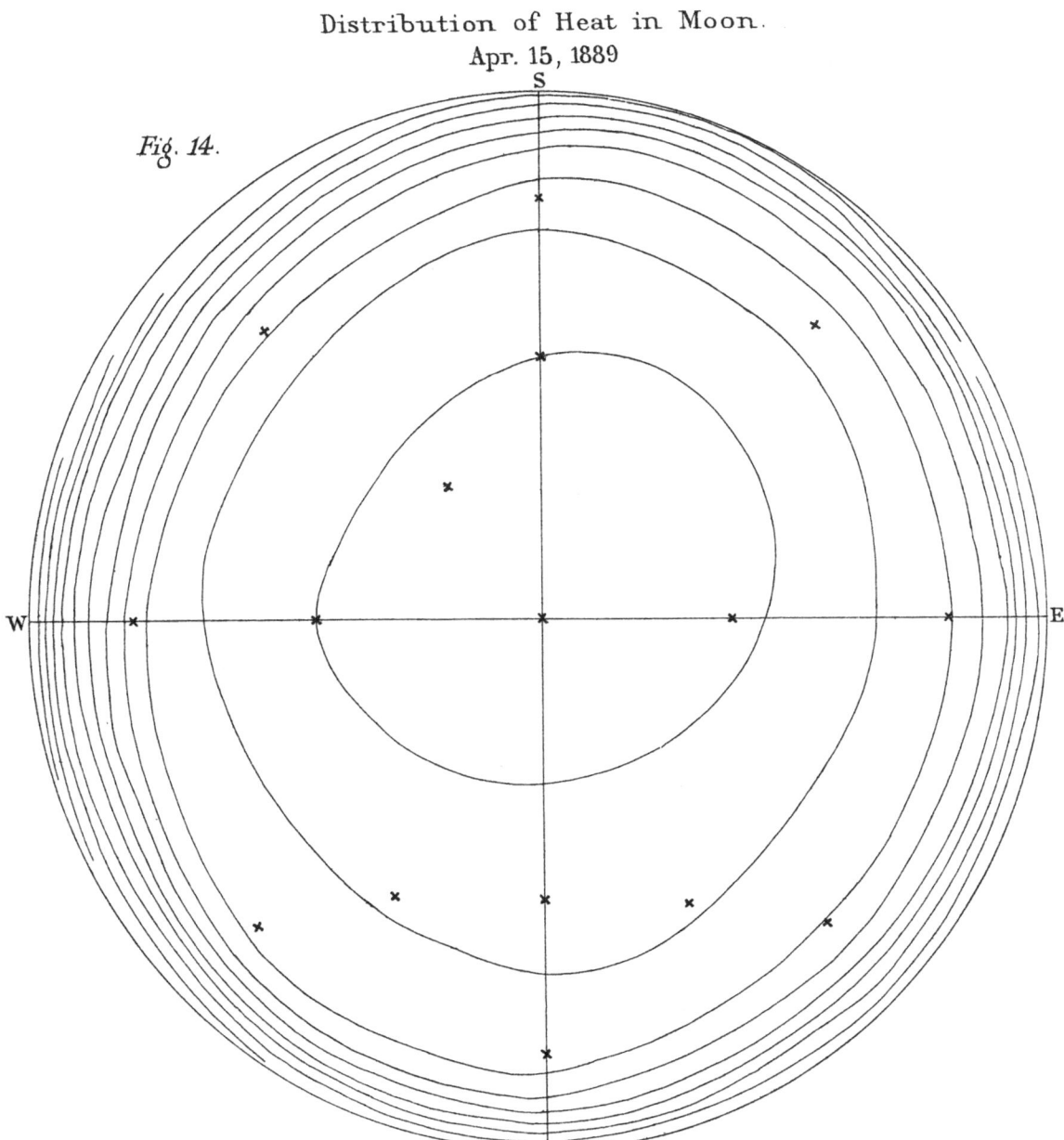

Distribution of Heat in Moon.
Apr. 15, 1889

Fig. 14.

Phase - Curve.

Fig. 15.

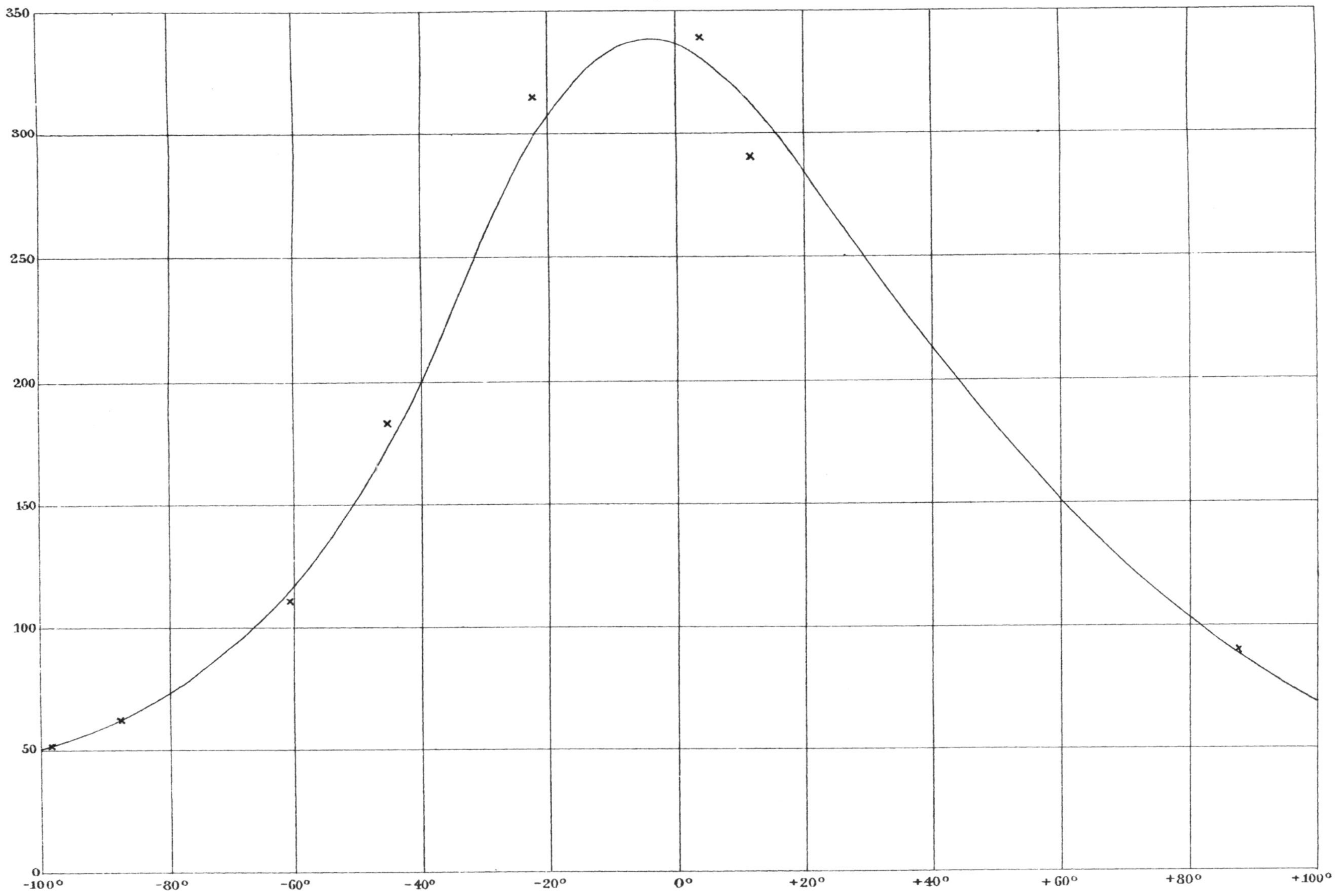